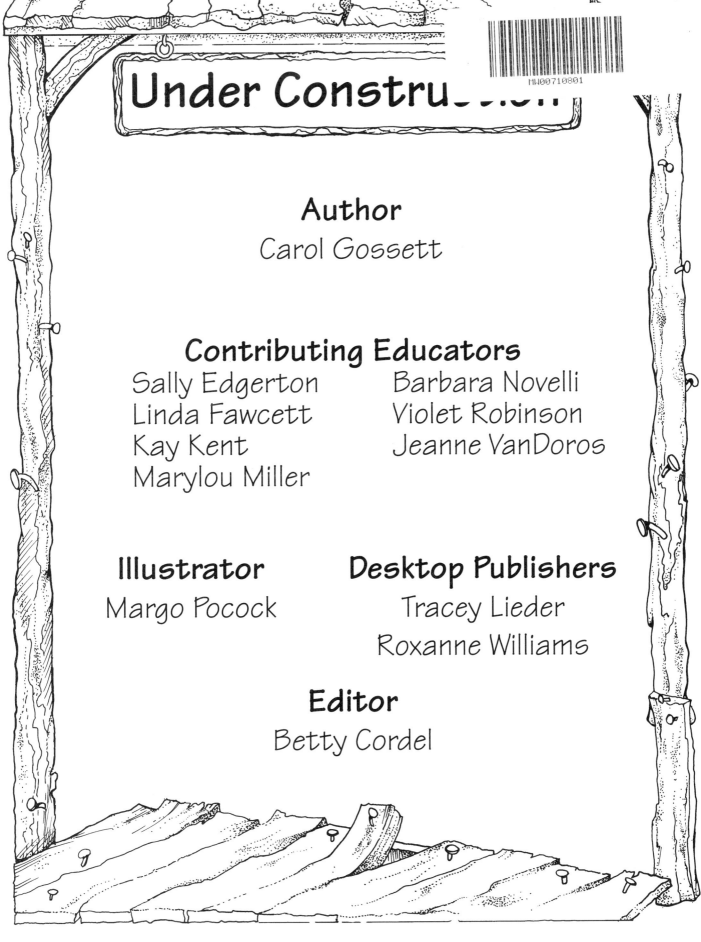

Under Construction

Author
Carol Gossett

Contributing Educators
Sally Edgerton
Linda Fawcett
Kay Kent
Marylou Miller

Barbara Novelli
Violet Robinson
Jeanne VanDoros

Illustrator
Margo Pocock

Desktop Publishers
Tracey Lieder
Roxanne Williams

Editor
Betty Cordel

Developed and Published
by
AIMS Education Foundation

This book contains materials developed by the AIMS Education Foundation. **AIMS** (**A**ctivities **I**ntegrating **M**athematics and **S**cience) began in 1981 with a grant from the National Science Foundation. The non-profit AIMS Education Foundation publishes hands-on instructional materials that build conceptual understanding. The foundation also sponsors a national program of professional development through which educators may gain expertise in teaching math and science.

AIMS Education Foundation
P.O. Box 8120, Fresno, CA 93747-8120 • 888.733.2467 • aimsedu.org

ISBN 978-1-881431-64-X

Printed in the United States of America

Table of Contents

I Hear and I Forget

I See and I Remember

I Do and I Understand

–Chinese Proverb

Under Construction
Introduction

Technology for young learners is the application of their understanding and use of materials, tools, and design. Throughout this study, they apply the use of technology by making constructions for themselves as well as for characters from child-centered literature. Young children can spend hours at home making structures with blocks, boxes, sand and water, snow, etc. The same is true in the classroom. By capitalizing on this high level of engagement, technological construction becomes the vehicle for facilitating the development of many concepts, inquiry skills, and appropriate vocabulary. In their collaboration with others, the students can work in a risk-free environment. In these experiences, "failures" are merely occasions to try again. Often as much can be learned from things that do not work as from things that do. When students compare and contrast their results, they will begin to observe and communicate cause-effect relationships that can be developed later into the forming of hypotheses.

The text guides in focusing the lesson
- by posing interesting problems to heighten curiosity;
- by asking thought-provoking questions about the materials, tools, and designs to stimulate thinking;
- by suggesting ways of testing and evaluating the constructions to help develop analytic skills; and
- by providing for open-ended exploration to nourish creativity.

The challenge for the teacher will not be to organize and explain complicated concepts dealing with the technology of construction, but to help children interact comfortably with one another and with the materials and tools provided as they begin to understand the design process. Ample time needs to be allotted for the students to thoroughly immerse themselves in their projects. Long-term projects encourage persistence, provide opportunities for children to process their learning, and allow them to build on information they gather as they continue to investigate and alter their designs. A variety of materials and space to store the on-going projects will add to the success of this study. Although the activities tend to be material-intensive, most materials are common household items which can be easily solicited from the students' families.

The activities broaden in scope as the students progress through them. They begin by studying isolated components of technology — materials, tools, and design — and then integrate them. This helps the students understand the interdependency of the three components. The study is culminated by inviting students to use the constructions they have made to put on puppet shows and retell stories using pop-up, "designer" books.

Under Construction should be used to encourage students to become life-long learners. One excited student in a Kindergarten classroom where some of the activities were being field tested helped to sum up her learning by saying, "I learned that you have to try things over and over and over to get them right." Truly, this is a valuable lesson for us all.

Under Construction
Overview

Technology

Through the study of technology, we apply the use of materials, tools, and design to change our world.

Materials

- Some things are made by nature and others are manufactured.
- Some materials are suited for certain purposes, others are not.

Tools

- Tools are used to help us observe, measure, and make things.
- Tools help us do things better or more easily.
- Some tools are useful for certain jobs, others are not.

Design

- Some designs are good for certain purposes, others are not.
- When designing something, the purpose of the project, the materials used, and the tools available must be considered.

Size and Scale

- The size of an object can be determined by comparing it to another object.
- Objects can be grouped or classified according to scale. They can be described using words such as small, smaller, smallest; big, bigger, biggest; etc.

Project 2061 Benchmarks*

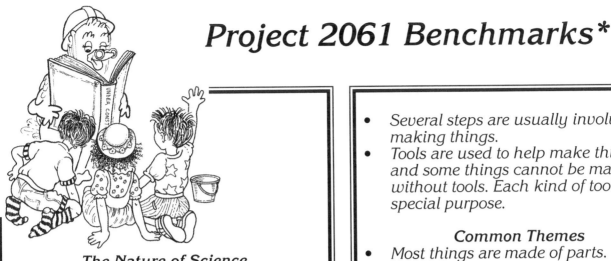

The Nature of Science
- Describing things as accurately as possible is important in science because it enables people to compare their observations with those of others.

The Nature of Mathematics
- Circles, squares, triangles, and other shapes can be found in things in nature and in things that people build.

The Nature of Technology
- Tools are used to do things better or more easily and to do some things that could not otherwise be done at all. In technology, tools are used to observe, measure, and make things.
- When trying to build something or get something to work better, it usually helps to follow directions if there are any or to ask someone who has done it before for suggestions.
- People can use objects and ways of doing things to solve problems.
- People may not be able to actually make or do everything that they can design.
- People, alone or in groups, are always inventing new ways to solve problems and get work done. The tools and ways of doing things that people have invented affect all aspects of life.

The Designed World
- Some kinds of materials are better than others for making any particular thing. Materials that are better in some ways (such as stronger or cheaper) may be worse in other ways (heavier or harder to cut).

- Several steps are usually involved in making things.
- Tools are used to help make things, and some things cannot be made at all without tools. Each kind of tool has a special purpose.

Common Themes
- Most things are made of parts.
- Something may not work if some of its parts are missing.
- When parts are put together, they can do things that they couldn't do by themselves.
- Many of the toys children play with are like real things only in some ways. They are not the same size, are missing many details, or are not able to do all of the same things.
- A model of something is different from the real thing but can be used to learn something about the real thing.
- One way to describe something is to say how it is like something else.

Habits of Mind
- Use hammers, screwdrivers, clamps, rulers, scissors, and hand lenses, and operate ordinary audio equipment.
- Assemble, describe, take apart and reassemble constructions using interlocking blocks, erector sets, and the like.
- Make something out of paper, cardboard, wood, plastic, metal, or existing objects that can actually be used to perform a task.

* American Association for the Advancement of Science. **Benchmarks for Science Literacy.** Oxford University Press. New York. 1993.

NRC Standards*

Abilities Necessary to do Scientific Inquiry

- Ask a question about objects, organisms and events in the environment.
- Plan and conduct a simple investigation.
- Employ simple equipment and tools to gather data and extend the senses.
- Use data to construct a reasonable explanation.
- Communicate investigations and explanations.
- Tools help scientists make better observations, measurements, and equipment for investigations. They help scientists see, measure, and do things that they could not otherwise see, measure, and do.

Understanding About Scientific Inquiry

- Scientists use different kinds of investigations depending on the questions they are trying to answer. Types of investigations include describing objects, events, and organisms; classifying them; and doing a fair test (experimenting).

Properties of Objects and Materials

- Objects have many observable properties, including size, weight, shape, and color. Those properties can be measured using tools, such as rulers and balances.

- Objects are made of one or more materials, such as paper, wood, and metal. Objects can be described by the properties of the materials from which they are made, and those properties can be used to separate or sort a group of objects or materials.

Understanding About Science and Technology

- People have always had problems and invented tools and techniques (ways of doing something) to solve problems.

Abilities of Technological Design

- Identify a simple problem.
- Propose a solution.
- Implement proposed solutions.
- Evaluate the design.
- Communicate the problem design and solution.

Abilities to Distinguish Between Natural Objects and Objects Made by Humans

- Some objects occur in nature; others have been designed and made by people to solve human problems and enhance the quality of life.
- Objects can be categorized into two groups, natural and designed.

* National Research Council. **National Science Education Standards.** National Academy Press. Washington, D. C. 1996.

NCTM Standards*

Mathematics as Problem Solving

- Formulate problems from everyday and mathematical situations
- Develop and apply strategies to solve a wide variety of problems

Mathematics as Communication

- Relate physical materials, pictures, and diagrams to mathematical ideas

Mathematical Connections

- Use mathematics in other curricular areas
- Use mathematics in their daily lives

Estimation

- Apply estimation in working with quantities, measurement, computation, and problem solving

Number Sense and Numeration

- Construct number meanings through real-world experiences and the use of physical materials
- Develop number sense
- Interpret the multiple uses of numbers encountered in the real world

Geometry and Spatial Sense

- Investigate and predict the results of combining, subdividing, and changing shapes
- Develop spatial sense
- Recognize and appreciate geometry in their world

Measurement

- Understand the attributes of length, capacity, weight, area, volume, time, temperature, and angle
- Develop the process of measuring and concepts related to units of measurement
- Make and use estimates of measurement

Statistics and Probability

- Collect, organize and describe data

*   National Council of Teachers of Mathematics. **Curriculum and Evaluation Standards for School Mathematics.** The National Council of Teachers of Mathematics, Inc. Reston, Virginia. 1989.

Materials

Made by Nature and Made by Me!

Topic
Technology: Materials

Key Question
What can we make with natural materials?

Focus
To help build an understanding of what manufactured objects are, the students will use natural materials to make something.

Guiding Documents
NRC Standards
- *Objects can be categorized into two groups, natural and designed.*
- *Some objects occur in nature; others have been designed and made by people to solve human problems and enhance the quality of life.*

NCTM Standard
- *Collect, organize, and describe data*

Math
Equalities and inequalities
Counting

Science
Technology
    natural and manufactured objects

Integrated Processes
Observing
Comparing and contrasting
Collecting and recording data
Communicating
Generalizing

Materials
For the class:
    a large graph (see *Management 1*)
    glue gun and glue sticks (see *Management 4*)

For each student:
    a rock (see *Management 2*)
    pin backs (see *Management 3*)
    object tag
    one or two 3" x 5" note cards

Background Information
Students can begin to build an understanding of natural and manufactured materials by participating in simple projects which give them experience in distinguishing the two groups.

In this activity, the students gather rocks which they will identify as natural, coming from Earth. They will then use these nature-made rocks to make jewelry. They become the manufacturers and thus experience first-hand the concept of manufactured objects. The students are then guided through a developmentally appropriate research project about other objects which are made from natural materials.

Management
1. Make a class graph without labels. The labels will be filled in during *Part One*.
2. Ask the students to collect 2-3 quarter-sized rocks or you may want to purchase small polished rocks for the students to use in their jewelry.
3. From a craft store or the craft section of a department store, purchase pin backs to attach to the rocks to make the jewelry.
4. Provide a strong glue such as the type used in cool temperature glue guns.
5. Duplicate one tag per student.
6. *Optional:* If possible, bring in a collection of jewelry made from natural materials.

Procedure
Part One
1. Discuss how natural objects are objects from nature; they come from the Earth.
2. Give each student a *Natural, Comes from Earth* tag and crayons. Take the class outside and ask them to place their tags on objects they think are nature-made or natural. Once students have placed their tags, have them describe to the class the object they chose to tag. Once everyone has shared their selections, direct the students to retrieve their tags and to draw a picture on the tag of the natural objects they chose.
3. Gather the students back in the classroom with their tags. Discuss the different things they tagged. Make graph labels for the different categories. Examples might be: trees, sticks, rocks, flowers, weeds, leaves, etc.
4. Direct the students to look at the graph and analyze the information with the following types of questions. Which natural object has the most tags? ... the fewest tags? Do any objects have the same number of tags? How many students tagged trees? How many more_____ are there than_____?

5. Using the graph for a reference, ask them to name some ways that people use these natural objects. [Responses will vary: flowers for decorations, trees to hold our swings or for shade, rocks to make fences, etc.]

Part Two
1. Using the collected rocks, ask the students where they think rocks come from.
2. Give each student a rock. Once the students understand that rocks come from the Earth and so are considered natural, ask them what they think they could make from their rock.
3. After several suggestions have been made, discuss with the students how people use natural objects to make things. If available, show the class your collection of jewelry made from natural objects.
4. Tell the students that they are going to make a piece of jewelry using their rocks.
5. With adult assistance, help students glue the rock to the pin back.

Part Three
1. Once the students have completed their jewelry, gather them together for a discussion about what they just experienced. Ask them to describe how they made their piece of jewelry. [We started with a natural object from the Earth, the rock, added something that was made by people, the pin back, and made a piece of jewelry]. Explain to them that this is what we call *manufacturing* a product. Something that is manufactured is made by people or by people using machines and tools. Tell them that we often start with something natural and then change it to make something new.
2. Invite the students to look around their classroom to find other things they think may have started with something natural and then was manufactured into a new object.
3. Ask the students to gather their tags used in *Part 1* and to place them on items in their room that they think may be made from these natural objects. (Some tags may not be used, while others might have several possibilities.)
4. Discuss their selections and assist the students with any misconceptions they may demonstrate.
5. Lead the class in a discussion about how things made by nature are not manufactured.

Part Four
1. Invite the students to be scientists and begin a research project. Give each student a 3″ x 5″ card. Direct them to find something in the classroom that they have a question about as to what it is made from. Examples might include: glass in a window or a mirror, plastic toy, crayon, chalkboard, bulletin board, book, etc. Direct the students to draw a picture of the item, then to ask someone to write the name of this item on their 3″ x 5″ card.
2. Give the students an opportunity to ask their parents, other teachers, the principal, librarian, older students, etc. for information about what the item is made from.

3. Once the students have found their answers, ask them to share their information with the class.
4. Display their note cards on a bulletin board.
5. Provide extra cards for the students who want to find out about additional objects in their environment.

Discussion
1. Name some objects that are natural.
2. What does it mean for an object to be natural?
3. Name some objects that are manufactured.
4. What does it mean for an object to be manufactured?
5. Name some natural materials which are used in manufactured objects.
6. Is there anything that you can think of that is not manufactured and not natural? Explain. (There is nothing in this category.)
7. Show the class manufactured objects which were made directly from natural materials. For example: woven baskets, jewelry, decorative rock figures, a wooden desk or chair. Ask students to describe the materials in the objects.
8. Are living things manufactured or natural? Explain why you think so.
9. Tell the class what you learned from your research project.
10. Describe something that you learned that was a surprise to you. (For example: glass is made from sand, paper from wood, cloth from cotton or wool, etc.)

Extensions
Reinforcement
1. Teach students to make other items from natural materials: use long pine needles to weave a basket, use long pieces of grass to make a friendship bracelet, turn a rock into a paperweight, etc.
2. Watch a video about the process of using natural materials to manufacture other objects. For example: the process of cotton to clothing, wool to cloth, trees to pencils, etc. *Wool Challenge* (available through Victorian Video Productions, 1-800-848-0284) takes the viewer through the process of shearing the sheep, processing the wool, spinning it, and weaving a jacket. The National Cotton Council offers a video, *From Fiber to Fabric,* that is similar, taking the viewer through the processing of cotton into fiber.

Enrichment
1. Take a field trip to a manufacturing plant. Some examples are: a paper plant, a cotton gin, a flour mill, a potato chip factory, etc.
2. Explore the things manufactured from natural materials by Native Americans and other cultures.

Made by Nature and Made by Me!

Natural
Comes from Earth

Natural
Comes from Earth

Natural
Comes from Earth

Natural
Comes from Earth

Natural
Comes from Earth

Natural
Comes from Earth

All Sorts of Stuff

Topic
Technology: Materials

Key Question
Where do we get materials?

Focus
The students will sort and classify a variety of materials.

Guiding Documents
Project 2061 Benchmark
- *Describing things as accurately as possible is important in science because it enables people to compare their observations with those of others.*

NRC Standards
- *Objects are made of one or more materials, such as paper, wood, and metal. Objects can be described by the properties of the materials from which they are made, and those properties can be used to separate or sort a group of objects or materials.*
- *Some objects occur in nature; others have been designed and made by people to solve human problems and enhance the quality of life.*

NCTM Standard
- *Collect, organize and describe data*

Science
Technologh
   characteristics of materials

Integrated Processes
Observing
Contrasting and comparing
Communicating

Materials
For the class:
   old magazines
   a variety of materials such as:
      cloth
      paper
      wood
      metal
      plastic
      Styrofoam
      cotton
      clay

For each group of four students:
   a recloseable plastic bag with a variety of materials
   Sorting Mat (see *Management 3*)

For each student:
   Family Letter
   Science Journal

Background Information
   Webster defines material as: "what a thing is, or may be, made of."
   Children experience materials daily. Through their observations students will begin to construct an understanding of the suitability of materials for serving a variety of purposes. Learning where the materials come from also increases the students' awareness of manufactured materials as compared to natural materials.

Management
1. It is suggested that you introduce the difference between natural and manufactured materials through the activity *Made by Nature and Made by Me!*
2. Two or three days prior to doing this activity, send home the *Family Letter* to generate a collection of materials from the students.
3. Duplicate and laminate (for reuse) one *Sorting Mat* for each group of students. If the materials brought in by students are large, you may need to enlarge the *Sorting Mats* or make your own.
4. *Part Two*: For each group of four students, place a variety of the materials from the exploration center into recloseable plastic bags. Add additional objects if you need more variety.
5. Copy one *Science Journal* per student.

Procedure
Part One
1. Once the students have begun to bring in a variety of materials, place these objects in an exploration center for at least a week.
2. After the entire class has had some experience in this center, bring them together for a class discussion.
3. Ask a student to choose one item from the center. Direct that student to describe the material.

4. Ask what they think the material is made from.
5. Ask the class if they think this material was manufactured or from the Earth, natural. (If appropriate ask them to describe how the manufactured items were made.)
6. Discuss the name for the material chosen.
7. Repeat the procedure with several different students.

Part Two
1. Divide the students into groups of four and give each group a bag of materials and a *Sorting Mat*. Invite them to sort these materials according to their own sorting rules.
2. Once all the groups have sorted their materials, ask them to share their sorting rules with the rest of the class.
3. Continue this procedure two more times.
4. Give each group a *Sorting Mat*. Tell them to put all the *manufactured* materials on one side and all the *natural* materials on the other. Ask the students to tell you why and how they made their decisions for where to place their materials.
5. Instruct the students to clear their *Sorting Mats*. Tell them you are going to change the rules. Now sort according to the following rules:
   Materials that *would be good to use to make clothes* on one side, and those that *would not be good to make clothes* on the other.
6. Direct the groups to report their results to the class.
7. Follow this same format for these suggestions:
   Materials that would be good to *build something* —and *not good for building*.
   Materials that would *keep things dry* if they were in water — and *not keep things dry*.
   Materials that are *strong* — and *not strong*.
   Materials made from *wood* — and *not wood*.
   Materials made from *paper* — and *not paper*.
   Materials made from *metal* — and *not metal*.
   Materials made from *plastic* — and *not plastic*.
8. After each of the sorting experiences above, ask the students to explain their choices of types of materials for the category being explored, discussing the similarities and differences between the materials the groups chose.
9. Talk about how some materials are suitable to use for several different purposes, while some are not suitable for certain purposes.

Part Three
1. Pass out a collection of magazines, drawing paper, crayons, and scissors to each of the students.
2. Give each student a *Science Journal*. Have them fold along the broken lines to make interior pockets as illustrated.
3. Direct the students to either cut out pictures from the magazines or draw pictures of things that are manufactured and others that are natural ... from the Earth.

4. Tell them to place these pictures in the appropriate pockets in their journals.

Discussion
1. Show your classmates something in the room made from wood. ... from plastic. ... from paper. ... from metal.
2. Chose a material from your collection. Tell your classmates what it is made from. Name some things that this material could be used to make.
3. Tell your classmates something you would like to make. Is there a material in your collection that would be good to use to make this object? Explain why or why not.
4. Show the class something in the room that is made by people. What material is it made with?
5. Where do we get materials?
6. Which of these materials are made by people?
7. What do we use materials for?

Extensions
1. Give the students an assignment to bring something from home which is made from a specific material: wood, paper, metal, or plastic.
2. Give the students an assignment to bring something from home which would be appropriate to use for a specific purpose: to store something, to wear, to play with, to eat with, etc.

All Sorts of Stuff

Dear Family,

Our class is learning about technology by observing and exploring materials, tools, and design.

Our first unit of study is about materials. Would you please help your child collect a variety of samples of materials to use in the classroom? These things will be shared with the rest of the class, so please do not send anything that needs to be returned. Here are some suggestions to help you:

cloth	Styrofoam
paper	cotton
wood	clay
metal	plastic

Thank you for your help with this collection. We will use these items to explore and discover many things in our study of technology.

All Sorts of Stuff
Sorting Mat

All Sorts of Stuff

Name _____

Science Journal

Materials Matter

Topic
Technology: Materials

Key Question
What material(s) will you use in your project?

Focus
Students will explore different materials and their uses.

Guiding Documents
Project 2061 Benchmark
- *Some kinds of materials are better than others for making any particular thing. Materials that are better in some ways (such as stronger or cheaper) may be worse in other ways (heavier or harder to cut).*

NRC Standard
- *Objects are made of one or more materials, such as paper, wood, and metal. Objects can be described by the properties of the materials from which they are made, and those properties can be used to separate or sort a group of objects or materials.*

Science
Technology
    materials

Integrated Processes
Observing
Contrasting and comparing
Communicating
Applying

Materials
For the class:
    a variety of materials: (see *Management 1*)
        cloth
        paper
        wood
        metal
        plastic
        Styrofoam
    Project Cards (see *Management 4*)

For each pair of students:
    Display Board Pieces (see *Management 2*)
    file folder (see *Management 3*)

Background Information
Once the young learners can identify different types of materials, they are ready to begin a study of their uses and suitability for different purposes. The students will discover through experience and discussion that some materials are appropriate for some purposes and inappropriate for others.

Management
1. At an exploration center, display the collection of materials from the activity *All Sorts of Stuff.* You will also need to add materials which would be appropriate to use with the *Project Cards.*
2. Duplicate one copy of the *Display Board Pieces* for each pair of students.
3. Construct a sample *Display Board* using a file folder and the *Display Board Pieces* provided.

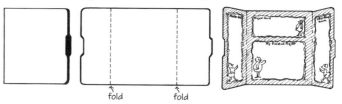

4. Duplicate one *Project Card* for each pair of students. Try not to have a pre-conceived idea of one correct response for each of these cards. There are several possibilities of materials to use with each of these projects. The intention of this activity is for the students to discover that there are multiple possibilities and that some choices of materials are better for some reasons and others are better for other reasons.

Procedure
Part One
1. Divide the class into partners. Hand out one *Project Card* to each set of partners.
2. Ask them to discuss the project and to choose a material from the exploration center they think would be best to use for the project. Tell them they will be reporting to the rest of the class when they are finished.
3. Allow time for students to gather their materials and to complete their projects.

Part Two

1. Explain to the students that they will now be preparing a *Display Board* to use when telling others about their projects. Show the students the sample board you have prepared.

2. Distribute the *Display Board* to each group. Tell the students to glue their *Project Card* on the appropriate display page. Demonstrate how to cut out this page and where to glue it onto their file folder *Display Board*.

3. Instruct the students to draw a picture of their project on the *My Finished Project* piece, and, if available, to glue a sample of the material they used on the *Materials Used* piece.

4. Direct them to draw pictures of other materials they could have used on the *Other Materials* piece.

Part Three

1. Have the students set up their *Display Boards* on a desk, table, or counter around the room. Tell them to place their completed projects in front of the boards.

2. Call on one pair of students at a time to give an oral presentation about their project and allow them to use the display board as a visual aid. Use the following questions to help generate a discussion with the students. Encourage the class to also ask the presenters questions about the projects.

Discussion

1. What was your project?
2. What material(s) did you use?
3. Why did you choose this material?
4. What are some things you needed to consider before making a decision on the best material to use?
5. Are there any other materials that might work for this project?
6. Which materials would not work with this project? Explain why.
7. Choose a material from the exploration center. Describe a project that this material would be good for.
8. Describe a job or a project you would like to do. Tell the class which materials would be good to use and describe how you would use these materials.

Extensions

1. Ask the students for ideas about other projects they would like to build. Suggest they explore two different materials for building their projects.

2. Direct the students to find two like objects that are made from different materials. Ask them to explain why they think the objects are made of different materials. Examples are: school binders (some are cloth covered cardboard others are plastic covered cardboard), balls (some are made of plastic and some are made of rubber).

Materials Matter

Project Cards

Write a letter to a friend.	Make a snack for the class.
Draw a picture for our class bulletin board.	Paint a picture.
Make a container to carry water.	Make a book.
Make a container to hold waste paper.	Make a chair for a stuffed animal.
Make a shirt for a stuffed animal.	Make a holder for your lunch money.
Make a container to hold your crayons.	Make a graph.
Make a sign for outside the classroom.	Make a holder to carry your school papers home.
	Wrap a box like a gift.

Display Board Pieces

Project Card

My Finished Project

Display Board Pieces

Materials Used

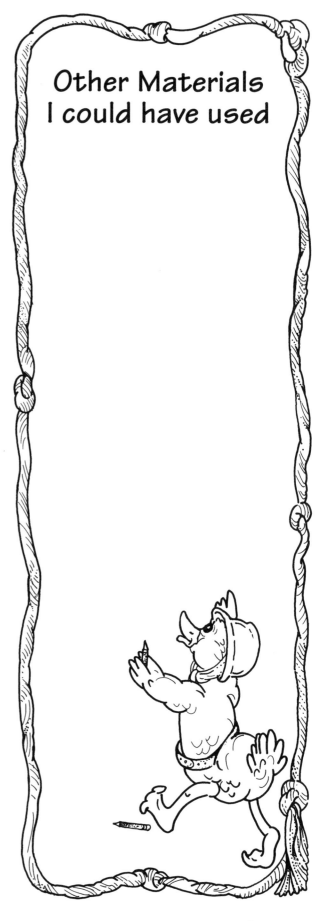

Other Materials
I could have used

14

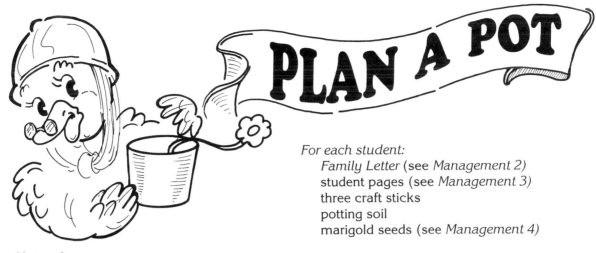

Topic
Technology: Materials

Key Questions
What material will you choose for making a planter?
Why did you choose this material?

Focus
The students will explore different materials and their suitability for use as a planter.

Guiding Documents
Project 2061 Benchmark
- *Some kinds of materials are better than others for making any particular thing. Materials that are better in some ways (such as stronger or cheaper) may be worse in other ways (heavier or harder to cut).*

NRC Standard
- *Objects are made of one or more materials, such as paper, wood, and metal. Objects can be described by the properties of the materials from which they are made, and those properties can be used to separate or sort a group of objects or materials.*

Science
Technology
   materials

Integrated Processes
Observing
Contrasting and comparing
Communicating
Applying

Materials
For the class:
   a variety of materials:
      cloth
      paper
      wood
      metal
      plastic
      Styrofoam

For each student:
   Family Letter (see *Management 2*)
   student pages (see *Management 3*)
   three craft sticks
   potting soil
   marigold seeds (see *Management 4*)

Background Information
This activity offers an opportunity for the students to construct their own understanding of the suitability of materials for different purposes through a trial-and-error experience. Young learners make connections in their understanding of these types of concepts through their successes as well as their "failures." They should be encouraged to think of their "failures" as "opportunities to try something else."

Management
1. At an exploration center, display the collection of materials from the activity *Materials Matter*.
2. Duplicate the *Family Letter* for each student.
3. Duplicate one set of flowers and labels for each student. Also, each student will need a copy of the flower pot.
4. Marigold seeds are suggested, however, most any annual flower seeds or grass seeds (rye grass sprouts quickly) would be appropriate.
5. Designate an area where the students can keep the plant containers for a couple of weeks where they will receive ample light and can be properly watered.
6. If some students fail to bring in materials, have them work with partners who did bring materials.

Procedure
Part One
1. Suggest to the students that you want to build a container to grow flowers outside the classroom. Ask them to discuss with a partner which material from the exploration center might be good for this project.
2. Ask the class to explain why they think the material they chose would be appropriate for this project.
3. Choose another material and ask them why they think their first choice would be better or worse or just as good for this project.
4. Tell the students to bring materials from home to build a plant container. Send the *Family Letter* home with the students to inform the parents of the assignment.

5. The following day, allow the students to make containers with the materials they chose and to plant and water the seeds.

6. Gather the class back together and discuss the materials each group used.

7. Discuss all the things they considered when choosing the appropriate materials to build the plant container. [needed to hold soil and water, needed to allow water to drain out the bottom, etc.]

8. Encourage the students to care for their seeds and to observe the condition of their containers for one or two weeks.

Part Two

1. Direct the students to look for any signs of damage to the container or the plant that may have occurred from the elements. Discuss the effects of the water, the weight of the soil, the sun, etc. on their containers.

2. Ask them to compare their container to two other containers in the room. Tell them to describe what may be better, the same, or unsatisfactory about the containers.

3. To record the results of their planter project, instruct the students to cut, fold, and glue a flower pot into a pocket-like holder. Direct them to cut out, glue and attach the labels: *soil*, *water*, and *sun* to the craft sticks.

4. To indicate whether or not the material they used was appropriate for the soil, the sun, and the water, tell them to use the happy-face and sad-face flowers by gluing the appropriate face onto each of their three sticks.

5. Instruct them to place these flower sticks into their pocket-like flower pots and to display them on a bulletin board.

Discussion

1. Describe your plant container and explain what type of material you chose and why.

2. What things did you have to consider before deciding which material to use?

3. What other materials do you think might work for this project?

4. Which materials would not work? Explain why.

5. What containers did not work? What material was used? Explain why you think it didn't work well. How can you make the container better?

6. Which materials worked as a planter? What material(s) was used to make it? Explain why you think this material was a good choice.

7. If you were going to build a plant container again, what materials would you want to use this time?

Extensions

1. Bring in different plant containers made from various materials such as clay, ceramic, plastic, wood, metal, cardboard (bio-degradable), and discuss their suitability in different growing conditions.

2. Invite a gardener to class to speak about different plant containers.

3. For further study or as a correlated unit of study, use the AIMS publication *Primarily Plants*.

PLAN A POT

Dear Family,

As we continue learning about technology, we are now studying the use of different materials. The assignment for tonight is for the students to find materials from home to use to build a container for a plant.

Please do not assist your child in the selection of material. We would like the students to construct their own knowledge about the materials through a trial-and-error process.

Thank you!

PLAN A POT

© 2007 AIMS Education Foundation

PLAN A POT

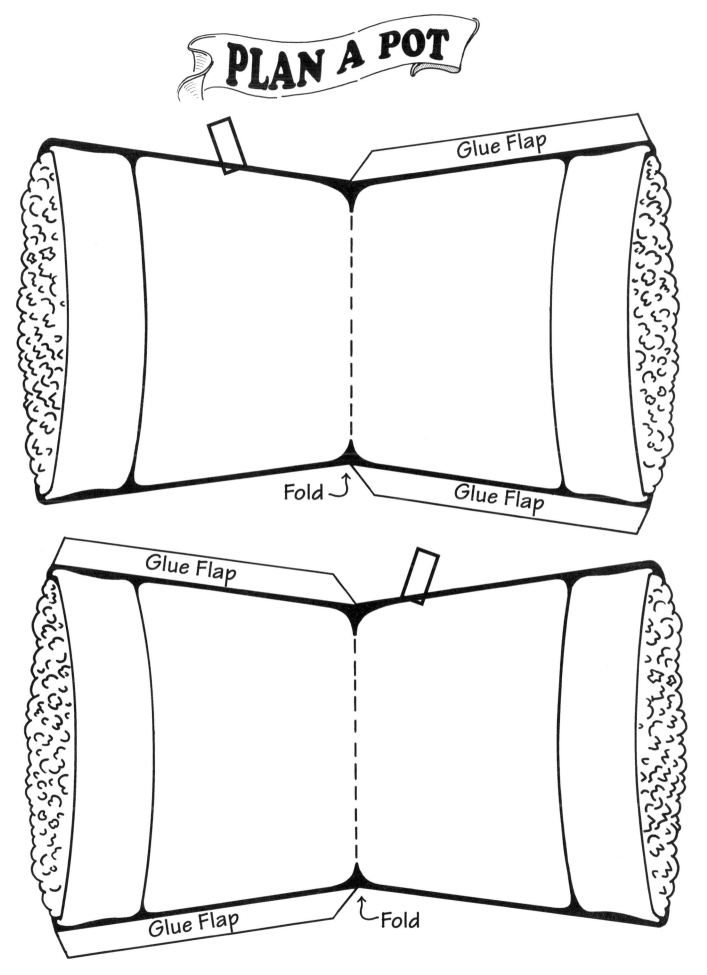

Glue Flap

Fold ↰

Glue Flap

Glue Flap

Glue Flap

Fold ↑

TALL WALLS

Topic
Technology: Materials

Key Question
What materials are appropriate to build a tall wall that will stay standing after the tennis ball jolt test?

Focus
Through trial-and-error experiences, the students will discover appropriate materials for building a "tall" wall.

Guiding Documents
Project 2061 Benchmarks
* *Assemble, describe, take apart and reassemble constructions using interlocking blocks, erector sets, and the like.*
* *Some kinds of materials are better than others for making any particular thing. Materials that are better in some ways (such as stronger or cheaper) may be worse in other ways (heavier or harder to cut).*

NRC Standard
* *Abilities of Technological Design:*
  * *Identify a simple problem.*
  * *Propose a solution.*
  * *Implement proposed solutions.*
  * *Evaluate the design.*
  * *Communicate the problem, design and solution*

NCTM Standard
* *Develop the process of measuring and concepts related to units of measurement*

Math
Measurement

Science
Technology
    materials

Integrated Processes
Observing
Communicating
Comparing and contrasting

Materials
For the class:
    a variety of building materials (see *Management 1*)
        wooden blocks
        clean, empty milk cartons
        cereal boxes
        cardboard boxes
        Lincoln Logs®
        LEGO® elements
        Unifix cubes
        Hex-a-link cubes
    large piece of cardboard (see *Management 2*)
    tennis ball

For each group:
    Brick Measuring Tape (see *Management 5*)

For each student:
    Tall Walls journal

Background Information
The children will learn as much from their failures as from their successes in this activity. In reconstructing and revising a structure that has not lived up to their expectations, children may try several different kinds of structural approaches before finding one that best suits their needs. They will borrow freely from their classmates, accepting, rejecting, and modifying their ideas as they see fit.

The students are challenged to build and build again until they find a design that creates the tallest possible wall which will remain standing after the base on which it is built is jolted by a tennis ball drop.

Through this activity, the students will begin to form an awareness of the building material including size, shape, and the effects of different forces on the materials. The students will begin to construct some basic design principles.

Management
1. Collect a large number of building blocks, milk cartons, cereal boxes, etc. Try building a wall or two with the materials you have available to determine the supplies necessary for your class.
2. Assign an area for each group of four students to work. Each group will need a flat surface such as a table top or a tiled floor area. It is suggested that a piece of cardboard serve as the base for the walls. A tennis ball dropped onto the cardboard will serve as the "jolt test" to determine the stability of the walls.

3. It is suggested the students work in groups of four in designated areas with the entire class participating at the same time.
4. The students should be given several opportunities for exploration of the building materials prior to introducing this lesson.
5. Duplicate and assemble the *Brick Measuring Tape* for each group.
6. Duplicate the *Tall Walls* journal back to back for each student.

Procedure

1. Review the nursery rhyme *Humpty Dumpty Sat On A Wall* with the students. Ask them from what materials they think Humpty's wall was made. Lead the students in a discussion of other materials that can be used to build a wall.
2. Challenge the students to build the highest possible wall using one type of building material that will be able to remain standing without any other type of support.
3. Demonstrate to the students how to do the tennis ball jolt test by simply dropping a tennis ball next to their structure. Discuss the possible results of the jolt test: the wall can *stay*, the wall can *shift*, or the wall can *spill*.
4. Divide the students into groups of four. Ask them to state the problem they are trying to solve. Direct them to discuss the problem with their classmates and to implement a proposed solution.
5. Allow time for the groups to build their walls.
6. Distribute a *Tall Walls* journal to each student and have them draw a picture of their *First Design*. Direct the students to perform the tennis ball jolt test on their walls and determine whether the wall stayed, shifted, or spilled.
7. Lead the class in a discussion as to the material used in their walls. Discuss why some seem to be more stable than others.

8. Allow the students to go back to their own walls and to continue building. Remind them to use the information they have gathered through their own construction, combined with what they have observed from their classmates, to build the tallest, sturdiest wall. Direct the students to perform the tennis ball jolt test on their walls.
9. After another building session, again compare and contrast the different materials used for the walls.
10. Have the students measure the walls by first ordering them from shortest to tallest and then by using the *Brick Measuring Tape*. For older students you may want to use a meter stick.
11. Have students illustrate their *Final Design* on the activity page.

Discussion

1. What was the problem you were trying to solve?
2. Describe how your group solved this problem.
3. Why were the materials you used successful? … unsuccessful?
4. What material did you use?
5. How did you discover what material worked best?
6. How did the shape of the material change the results?
7. What will you do differently next time you build a wall? … the same?
8. What other type of building materials might have worked better? Why?
9. Look at the different walls. Which ones seem to shift and spill more than others? What materials were used to build these walls?
10. Which ones seem to shift but still remain standing? What materials were used in these walls?
11. Which materials survived the most tests? Why do you think they survived?

Extensions

1. Leave the building materials out on a discovery table for the students to explore further during a free choice period.
2. Allow students to design other *tests* to determine the stability of their walls.

Brick Measuring Tape

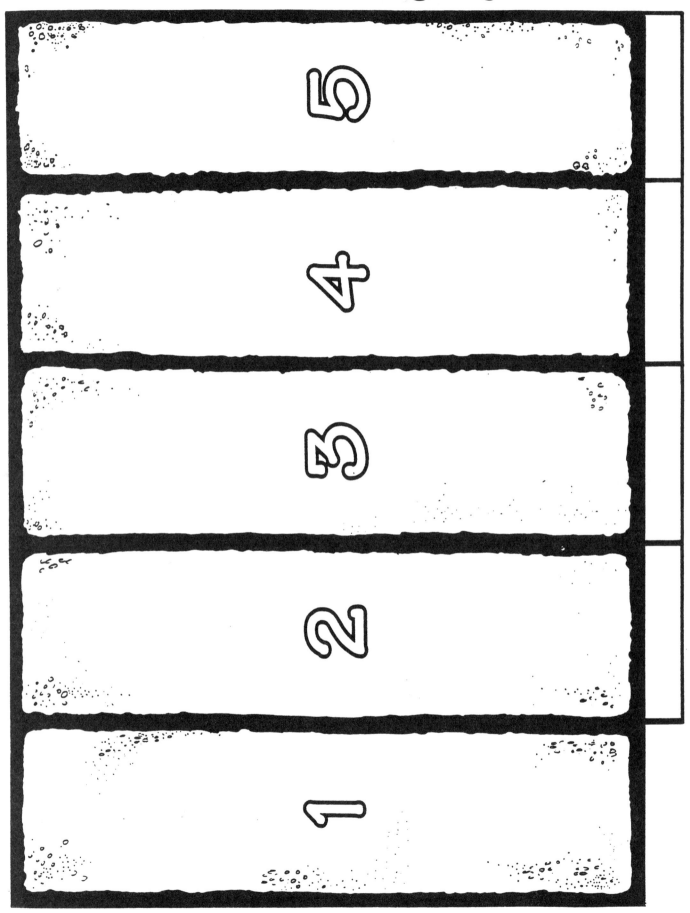

TALL WALLS

We discovered that ...

Next time I would like to ...

Our First Design

This wall was _____ bricks high.

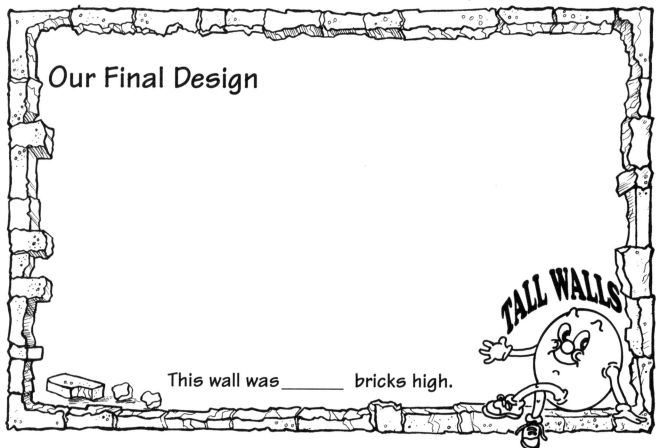

Our Final Design

This wall was _____ bricks high.

A Safe Landing

Topic
Technology: Materials

Key Question
What materials can protect a hard-boiled egg from cracking when dropped from a height of five "bricks"?

Focus
The students will explore different materials and their effectiveness in absorbing the energy of a hard-boiled egg which is dropped from a determined height.

Guiding Documents
Project 2061 Benchmark
• *Objects have many observable properties, including size, weight, shape, color, temperature, and the ability to react with other substances.*

NRC Standard
• *Abilities of Technological Design:*
  • *Identify a simple problem.*
  • *Propose a solution.*
  • *Implement proposed solutions.*
  • *Evaluate the design.*
  • *Communicate the problem, design and solution.*

NCTM Standard
• *Develop the process of measuring and concepts related to units of measurement*

Math
Charting
Measuring

Science
Technology
  materials

Integrated Processes
Observing
Contrasting and comparing
Collecting and recording data
Interpreting data
Communicating
Predicting

Materials
For the class:
  small hard-boiled eggs (see *Management 3*)
  class chart (see *Management 5*)
  materials for the landing pad (see *Management 9*)
    paper towels
    foam rubber
    newspaper
    Styrofoam packing material

For each student:
  Egg Hanger (see *Management 4*)
  30 cm (12") narrow ribbon or string
  journal, optional (see *Management 11*)

For each group of students:
  materials for walls (see *Management 2*)
  Brick Measuring Tape from *Tall Walls*
  cardboard box (see *Management 7* and *8*)

Background Information
Students will use different materials to try to help cushion the impact of a hard-boiled egg. When the egg falls, both the landing surface material and the egg are absorbing energy built up from the fall. When the egg hits the landing material, the material absorbs some of the egg's energy. Depending on the material used on the landing surface, the egg will crack if that material cannot absorb enough of the egg's energy. If the material absorbs enough energy, the egg will not crack.

The students will be observing these effects through participation in an investigation where they will be making a prediction to try to answer a question, planning and conducting a test, and drawing conclusions based on the results of these tests.

The young learners will begin to experience the simple concept that different materials are better than others to use as a landing pad to protect objects from a fall.

Management
1. Divide the students into groups of four.
2. Build walls similar to those built in *Tall Walls*. The walls should measure five "bricks" high or be elevated to that height.

3. Bring in at least two **small, cooled, hard-boiled** eggs per group of students in your class. More trials will require additional eggs.
4. Using cardstock, duplicate the *Egg Hanger* back-to-back, one per student.
5. Prepare a class chart as illustrated.

Did the egg break?

yes *no*

6. Use the *Brick Measuring Tape* from *Tall Walls*.
7. The landing surfaces should be placed inside a cardboard box to help contain the egg which may bounce upon landing.
8. Use cardboard boxes which are approximately 45 cm (18″) high, 45 cm (18″) wide, and at least 45cm (18″) long to hold the landing pad material. In order to standardize the thickness of the landing materials used, mark "fill" lines inside the cardboard boxes. For example, if a 7-8 cm (3″) depth of Styrofoam is used, then that same thickness of newspapers should also be used.
9. Materials listed can be provided for the students or you can present them with the challenge of providing a surface that will prevent the egg from cracking when dropped from a height of five "bricks." Students can then brainstorm materials and bring them from home for testing.

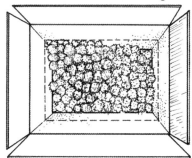

10. If time allows, students can use markers to draw faces on the eggs they will be dropping.
11. If desired for older students, duplicate the two page journal to use to record the materials, procedure, and results of their investigations. To conserve paper, copy the journal pages front to back.

Procedure

Part One

1. Review the nursery rhyme *Humpty Dumpty Sat On A Wall*. Discuss what might have been done to prevent Humpty from breaking. Lead the students into a discussion about possible materials to use as a landing surface.

2. In groups of four, challenge the students to build walls similar to those built in *Tall Walls*. Ask the students to place one egg on top of their wall, and then to gently push it off, allowing it to land on the floor or table top.
3. Compare the damage incurred by the eggs and discuss the landing surface.
4. Brainstorm possible materials that could be used to build a landing pad for the eggs.
5. Show the students the cardboard boxes which they will be using to hold the landing pads. Explain to them that they need to use these boxes to help contain the egg once it falls.
6. Explain to them the need for a *fill line* to help insure a *fair test*. Show the students the fill line and tell them they will need to gather enough materials for their box to reach that line.
7. Challenge the groups to build a landing surface that will protect the eggs. Tell them that they will be working as scientists to discover the answer to the following question: What is a good material to use to protect a hard-boiled egg from cracking when dropped from a height of five bricks?
8. Have the students state the problem they are trying to solve, discuss the problem with their group, and arrive at a possible solution.
9. Direct the students to collect the needed materials from home. (If this presents a problem, the teacher will need to bring in materials.)

Part Two

1. Direct the students to add their landing pad materials to their box up to the *fill line*.
2. On the front side of their *Egg Hanger*, instruct the students to glue a sample of the landing surface material they predict will allow the egg to survive this fall. Direct them to draw a picture of their own egg falling into this material or use the *Egg Cutouts* provided.

3. Once all the landing surfaces have been placed in the boxes, gather the students around for the egg test. Direct each group to appoint a *King*, a *Builder*, a *Checker*, and a *Recorder*. Direct the *Builder* to check that the material reaches to the fill line. Have the *Checker* measure the wall to check that it is five bricks high. Ask the class to recite the *Humpty Dumpty* nursery rhyme. When they finish, instruct the *King* to hold the hard-boiled egg on top of the wall and gently push it off.

4. Ask the Recorder of each team to mark the appropriate space on the *Did Your Egg Break?* class chart.

5. Continue these procedures for each landing pad that was built.

6. If several of the eggs break, send the students back to work with their groups to redesign their landing pads. Bring the class back for new trials.

7. On the back side of their *Egg Hangers*, direct the students to record the results of the tests on the different materials. Have them glue bits of each landing surface material tested to their chart.

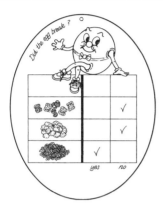

8. Punch a hole at the top of the egg hanger and tie a ribbon through it. If possible, hang them around the room so all can observe the different materials that were tested.

9. Direct the groups to discuss their solutions and to evaluate whether or not their choice of landing pad material was successful. If it needs to be improved, have them discuss how to make those changes.

10. If appropriate, provide the science journals to record their investigations.

Discussion

1. What problem did you need to solve?
2. Describe the plan your group designed.
3. Explain what happened when you tested your landing pad materials?
4. How was your test a "fair test"?
5. What was the height of your wall?
6. Using our class chart, which materials seemed to work best to help the eggs survive? ... work the least?
7. What do you notice about the materials that helped the eggs survive the fall?
8. What do you notice about the materials used that did not help the eggs survive?
9. What other materials do you think might work as well or better than the ones we used?
10. How much higher do you think we can drop our eggs and still have them survive on our best landing pads?
11. Show several different materials to the students and direct them to order them from the most likely to the least likely to help the egg survive a fall. Ask them to explain why they ordered them in this way.

Extensions

1. Allow the students to continue investigating additional materials.
2. Once the students have tried several materials and found a material or two that allows Humpty to survive the fall, suggest to them that they try a higher drop! Allow them to continue testing their material at different heights. Make sure to supervise or perform the egg drop when the height gets too high for the students to safely reach.

Egg Hanger — Front

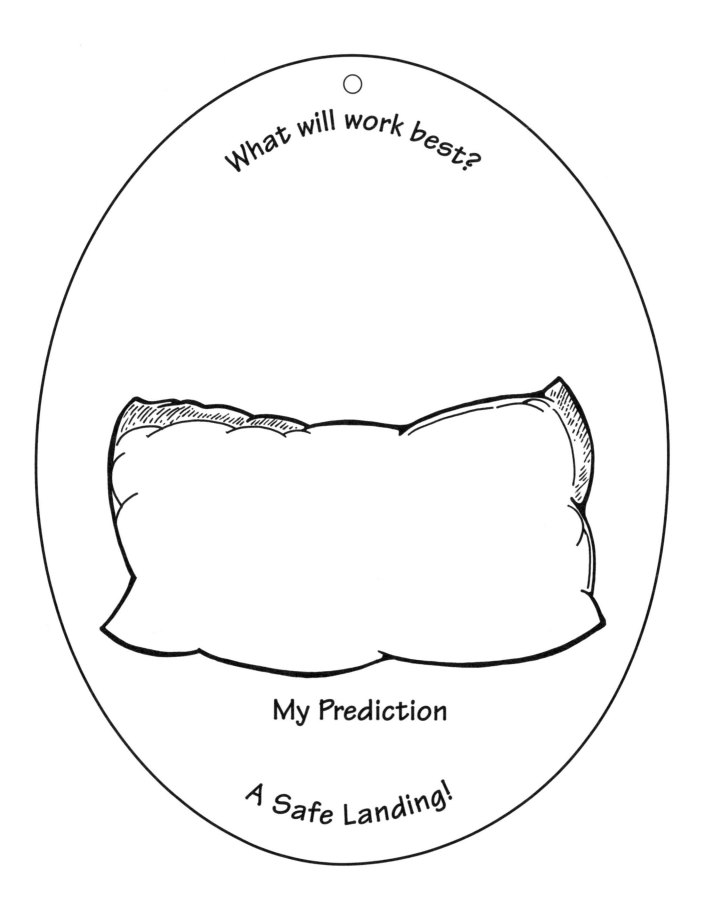

What will work best?

My Prediction

A Safe Landing!

Egg Hanger — Back

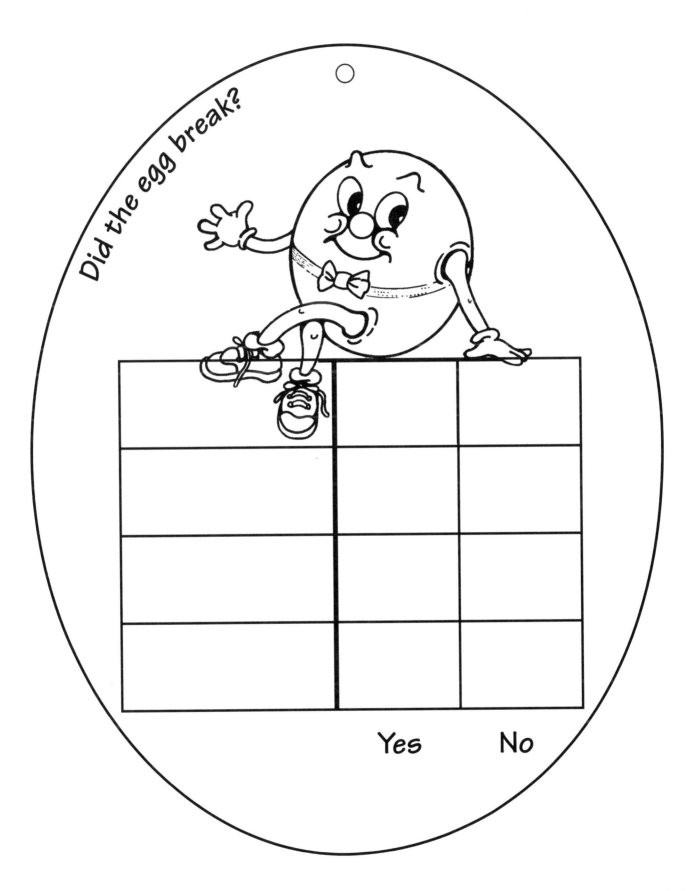

Did the egg break?

Yes   No

Egg Cutouts

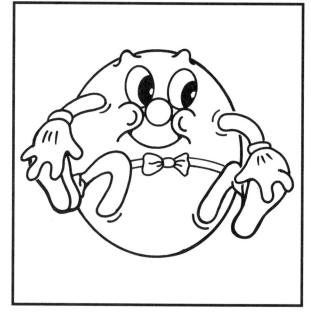

I learned

A Safe Landing

Materials used:

Results:

Did the egg break?

Test 1

Yes   No

What we did:

Test 2

Yes   No

Test 3

Yes   No

Test 4

Yes   No

Keep Trying

Words by Suzy Gazlay

Tune: Boom, Boom, Ain't It Great to be Crazy

To build a proj-ect we should choose, The

best ma-te-ri-als we can use. Pick

one to try, then an-oth-er to see What

makes it the best that it can be.

You know, we just have to keep try-ing; You

know, we just have to keep trying:

Build it a-gain out of some-thing else! You

know, we just have to keep try-ing.

Tools

Time for Tools

Topic
Technology: Tools

Key Question
What is a tool and how is it used?

Focus
The students will explore different objects which can be used as a tool.

Guiding Documents
Project 2061 Benchmark
* *Tools are used to do things better or more easily and to do some things that could not otherwise be done at all. In technology, tools are used to observe, measure, and make things.*

NRC Standard
* *Tools help scientists make better observations, measurements, and equipment for investigations. They help scientists see, measure, and do things that they could not otherwise see, measure, and do.*

Science
Technology
  tools

Integrated Processes
Observing
Contrasting and comparing
Communicating

Materials
For the class:
  magazines, optional (see *Management 4*)
  a variety of classroom objects that can be used as tools (see *Management 1*):
  scissors
  balances
  paper clips
  glue
  transparent tape
  stapler and staples
  ruler

For each group of students:
  Tool Box (see *Management 2*)

For each student:
  a student *My Box of Tools* book

Background Information
A tool is something people can use to do things and to solve practical problems. A study of tools following the investigations with materials is a natural sequence for early explorations of technology.

Primary students come to school with many experiences using technology. They operate bicycles, ride in many different forms of transportation, use kitchen utensils, operate a VCR and TV, use a brush and comb to care for their hair, etc. Children are naturally curious and excited to explore the use of the tools. These activities will focus on the identification of a variety of objects that can be classified as tools, their many uses, and problems these tools may have been designed to solve.

Management
1. Prior to beginning this lesson, duplicate and send home the *Family Letter* asking for various items to use as tools. Once the tools have been gathered, place them in an exploration center for three or four days.
2. Prepare a tool box for each group of four students in your classroom or provide the instruction page to older students to prepare their own tool boxes. Place a tool in each tool box different from the tools you have displayed at the exploration center. Examples might be: a turkey baster, salad tongs, a plastic eyedropper, a paper clip, transparent tape, ruler, or a can opener.
3. For each student, duplicate and fold a *My Box of Tools* book.
4. Collect a variety of magazines for the children to use when searching for pictures to cut out for their book.

Procedure
Part One
1. Once the entire class has had some experience at the exploration center, bring them together for a class discussion.
2. Ask a student to choose one item from the center that might be considered a tool. Ask "Do you think this item is a tool? Explain how you think the object is used."
3. Ask the class to think of other ways this item might also be used.
4. Repeat this procedure with other tools.
5. Ask the children if they think there are any objects in the exploration center that are not tools. If they

indicate that there are, ask a student to bring that item to the group.

6. Discuss why they think or don't think this item is a tool. Ask for suggestions as to what this item is used for.

7. Discuss with the students how a tool is anything that helps someone do a job, a project, work, etc. Ask them to name tools that a painter may use. ... a teacher. ... a student.

Part Two

1. Divide the class into groups of four. Give each group a tool box.

2. Tell the groups that you have placed a tool in each tool box. Ask them to discuss the tool and to try to think of as many uses as they can for this object. Tell them they will be reporting to the rest of the class when they are finished.

3. After the groups have had time to discuss the object in their tool box, gather them together for the group reports.

4. Ask the students to describe the tool and then to explain how they think the tool is used.

5. Encourage them to give more than one example of how this tool could be used to help with a job or a project.

Part Three

1. Direct each group to place another item from the classroom that they think is a tool into their tool boxes. Gather the students together for a class discussion.

2. Invite the students to play a guessing game in which they try to guess what is in each group's tool box. Set up the rules of the game: They may only ask questions that would give a yes/no answer. The object of the game is to try to guess the contents using as few questions as possible. Encourage the

students to ask questions such as: Is it used to build something? Is it used to cut something? Is it used in the kitchen? Does it hold things together? Is it used to measure something?

3. As an assessment of the student's understanding of tools, direct the students to either draw or collect pictures of tools from magazines to glue into their *My Box of Tools* books.

4. Gather the class together to share and discuss their *My Box of Tools* books.

Discussion

1. Name some things that we have in the exploration center that are tools. Why do you think these items are tools?

2. How are these tools used?

3. How could you use these tools for a different purpose?

4. If there were some items that we named as tools that you were surprised about, name them. Why do you think of them as tools now?

5. Choose a tool from the exploration center. Show the class how to use this tool safely.

6. Describe a job, a project, or some work you would like to do. Which tool (or tools) would you need to use and how you would use it (them)?

7. What type of problems do you think some of these tools were designed to solve?

8. Choose two pictures from your *My Box of Tools* book and describe a job these tools could help you with.

Extension

Collect tools from various cultures. Some examples might be chopsticks, different containers for cooking, etc. Share the name of the tools, how to use these tools, and ask for other suggestions of how the students think they could use these tools.

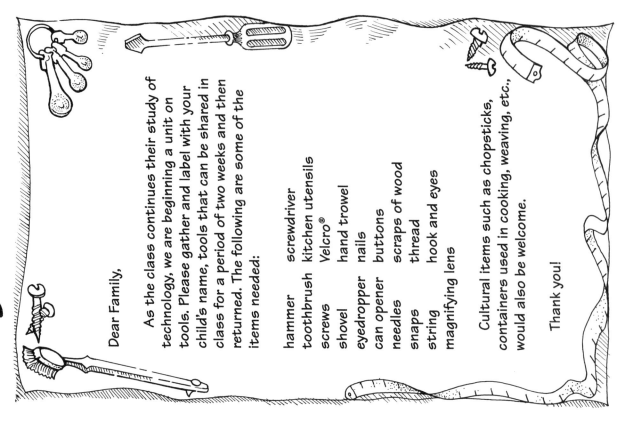

Time for Tools Family Letter

Dear Family,

As the class continues their study of technology, we are beginning a unit on tools. Please gather and label with your child's name, tools that can be shared in class for a period of two weeks and then returned. The following are some of the items needed:

hammer	screwdriver
toothbrush	kitchen utensils
screws	Velcro®
shovel	hand trowel
eyedropper	nails
can opener	buttons
needles	scraps of wood
snaps	thread
string	hook and eyes
magnifying lens	

Cultural items such as chopsticks, containers used in cooking, weaving, etc., would also be welcome.

Thank you!

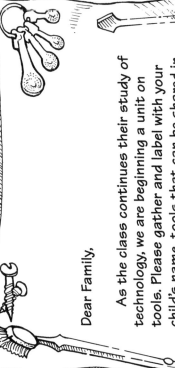

Time for Tools Family Letter

Dear Family,

As the class continues their study of technology, we are beginning a unit on tools. Please gather and label with your child's name, tools that can be shared in class for a period of two weeks and then returned. The following are some of the items needed:

hammer	screwdriver
toothbrush	kitchen utensils
screws	Velcro®
shovel	hand trowel
eyedropper	nails
can opener	buttons
needles	scraps of wood
snaps	thread
string	hook and eyes
magnifying lens	

Cultural items such as chopsticks, containers used in cooking, weaving, etc., would also be welcome.

Thank you!

Time for Tools

SHOE BOX LID

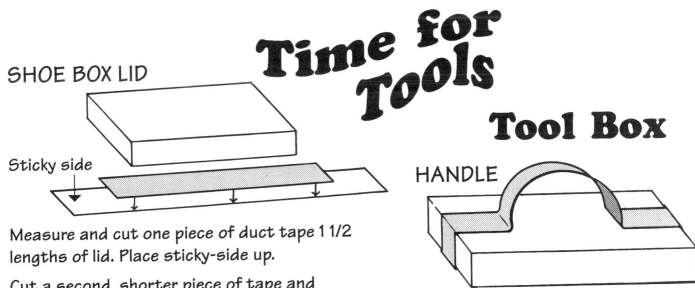

Sticky side

Measure and cut one piece of duct tape 1 1/2 lengths of lid. Place sticky-side up.

Cut a second, shorter piece of tape and center over longer piece, sticky-side down.

Tool Box

HANDLE

Form a handle, pressing sticky ends onto lid and wrapping around the edges of the box.

Cut a 20 cm piece of duct tape. Fold in half, leaving the end pieces unattached.

LATCH PIECE

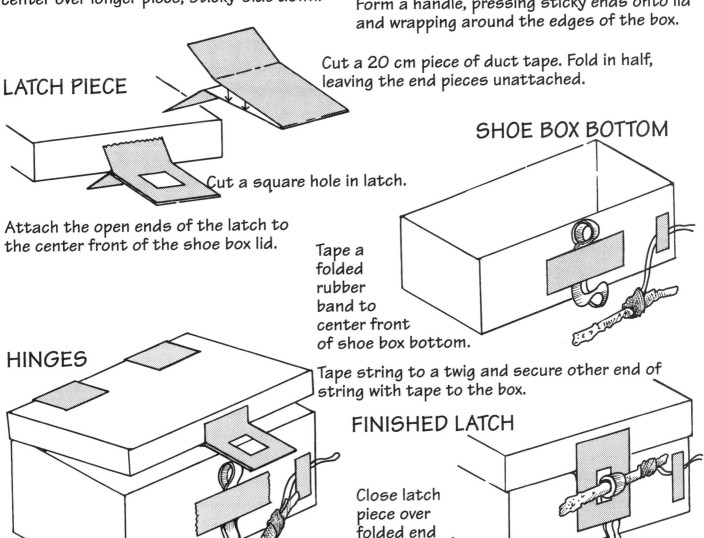

Cut a square hole in latch.

Attach the open ends of the latch to the center front of the shoe box lid.

SHOE BOX BOTTOM

Tape a folded rubber band to center front of shoe box bottom.

Tape string to a twig and secure other end of string with tape to the box.

HINGES

FINISHED LATCH

Close latch piece over folded end of rubber band. Secure by sticking the twig through the rubber band.

Use two pieces of duct tape to hinge the back side of the shoe box lid to the shoe box bottom.

My Box of Tools

_____

Name

Tools of the Trade

Topic
Technology: Tools

Key Question
How do you use tools?

Focus
The students will explore different tools and be given problems to solve using the tools.

Guiding Documents
Project 2061 Benchmarks
- Tools are used to do things better or more easily and to do some things that could not otherwise be done at all. In technology, tools are used to observe, measure, and make things.
- Tools are used to help make things, and some things cannot be made at all without tools. Each kind of tool has a special purpose.
- Tools such as thermometers, magnifiers, rulers, or balances often give more information about things than can be obtained by just observing things without their help.

NRC Standard
- Tools help scientists make better observations, measurements, and equipment for investigations. They help scientists see, measure, and do things that they could not otherwise see, measure, and do.

NCTM Standard
- Develop the process of measuring and concepts related to units of measurement

Math
Measuring
Using logical thinking

Science
Technology
    tools

Integrated Processes
Observing
Contrasting and comparing
Communicating
Applying

Materials
For the class:
    a variety of tools (see *Management 1 and 3*)

For each group:
    tool box (see *Management 2*)
    Challenge Card (see *Management 4*)
    Our Tool Project recording page (see *Management 4*)
    crayons
    scissors
    glue

Background Information
    To begin to understand technology in the world, a young learner needs to use different tools to solve practical problems.
    The students will work in small groups to decide which tools they will need to solve a problem and how those tools will help them with their challenge.

Management
1. Duplicate the *Challenge Cards* and gather the materials needed for each card. A caution to the teacher: Try not to have a pre-conceived idea of one correct response to these cards as there are several possibilities of tools to use with each of these projects. The intention of this activity is for the students to discover that there are multiple possibilities and that some choices of tools are better for some purposes than others.
2. Use the same tool box that was used in the activity *Time For Tools*.
3. Continue to use the tool collection from the exploration center in the activity *Time For Tools*.
4. Each group of students will need one-half of the *Our Tool Project* recording page.
5. Young primary students will need to work in pairs while older students will enjoy working in groups of four.

Procedure
1. Review with the students what tools are and ask for examples from the class.
2. Show a tool from the exploration center and ask the students for suggestions as to what that tool can help them do. Ask them to name multiple uses for the tool.

3. Provide each group with one of the suggested projects found on the *Challenge Cards*. Ask them to decide which tool they think would be best to use to solve the stated problem.

4. Allow each group time to discuss their plans, and to try out their ideas by solving the challenge.

5. Using their *Our Tool Project* recording page, instruct the students to glue their *Challenge Card* in the appropriate space and to draw pictures of the tools their groups used. Direct them to place the recording page and the completed projects in their tool boxes.

6. Ask the groups to report to the class by giving an oral presentation using the items in their tool box. (A suggestion is to ask the students to present a mock TV commercial in which they tell about the many uses for their tools.)

Discussion

1. Tell the class what your challenge was.
2. Describe the tool or tools you used.
3. Why did you choose this tool?
4. How did you use the tool to solve your problem?
5. What did the tools help you to do?
6. Was it a good tool to choose? What other tool could you have used?

Extensions

1. Ask the students to interview adults in their family to find out what types of tools they use in their jobs. Ask the students to share their findings with the class.

2. Invite several different people from the community to speak to the students and to show the students the tools they use in their jobs.

Tools of the Trade

Challenge Cards

Choose two pieces of wood. Use a tool or tools to attach the two pieces of wood together.

Take two pieces of paper and use a tool to attach them together.

Take a cloth and use a tool to attach a button to the cloth.

Draw a picture and then use a tool to hang it on the wall.

Find some paper. Use a tool to cut it.

Help yourself to some cereal. Use a tool to eat it.

Take a paper sack, put three blocks in it and use a tool to make it stay closed—even when the sack is turned upside down.

Make a Unifix cube train. Use a tool to measure how long it is.

Choose a rock. Use a tool to measure the mass of the rock.

Find a rock. Use a tool to look very closely at it. Use a tool to record all the colors you see on the rock.

Use a tool to open a can of olives.

Use a tool to lift a heavy box.

Use a tool to dig a hole in the sandbox.

Use a tool to clean the floor in the classroom.

Get some paint. Use a tool to paint a picture.

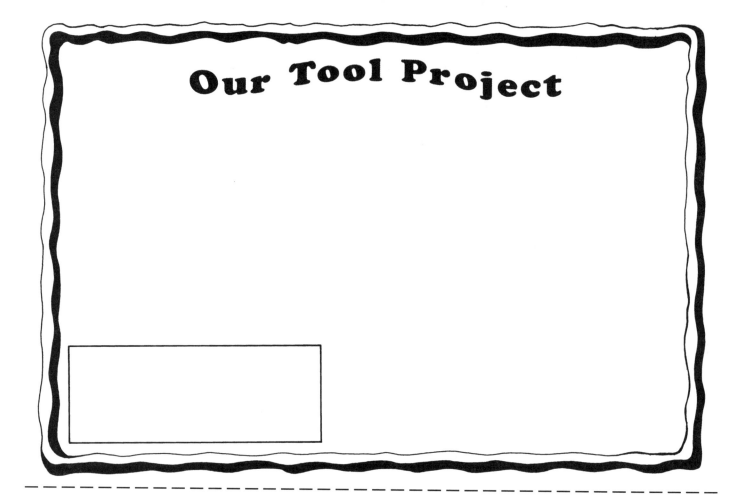

Our Tool Project

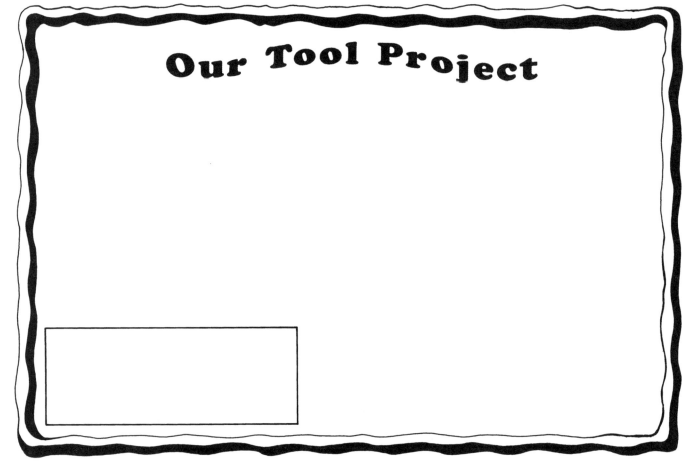

Our Tool Project

44

Tool Tales

Topic
Technology: Tools

Key Question
What tools were used in the nursery rhymes?

Focus
The students will gain an awareness of different tools which were used in nursery rhymes.

Guiding Document
Project 2061 Benchmark
- *Tools are used to do things better or more easily and to do some things that could not otherwise be done at all. In technology, tools are used to observe, measure, and make things.*

Science
Technology
    tools

Integrated Processes
Observing
Contrasting and comparing
Communicating

Materials
For the class:
    classic versions of nursery rhymes (see *Management 1*)

For each group of students:
    Nursery Rhyme Cards and *Questions* (see *Management 2*)
    one paper bag (see *Management 2*)
    five or more Unifix cubes per student (see *Management 4*)

For each student:
    one set of *Tool Cards* (see *Management 5*)
    one set of *Nursery Rhyme Cards* (see *Management 5*)
    large sheet of construction paper
    glue
    scissors

Background Information
Linking the information the students have been gathering about tools and their uses to nursery rhymes can serve as a highly motivating lesson. The students will choose cards with a picture representing the nursery rhyme and a question to answer regarding the possible tool that was used in the rhyme. This activity allows the students to use divergent thinking skills by applying and inferring information they have previously learned about the use of tools.

Management
1. The class should be familiar with the following nursery rhymes prior to beginning this activity. (See *Curriculum Correlation, Literature*)
    Jack and Jill
    Mary, Mary, Quite Contrary
    Three Little Kittens
    Little Miss Muffet
    Hickory Dickory Dock
    Little Boy Blue
    Jack Be Nimble
    Little Jack Horner
    Baa Baa Black Sheep

2. Duplicate on cardstock and laminate (for extended use) a set of *Nursery Rhyme Cards* and *Nursery Rhyme Questions* for each group of students. Be sure to duplicate on both sides, the nursery rhyme pictures on one side and the corresponding questions on the other. Cut out and place these cards into a paper bag.
3. For young learners who cannot read, plan to provide a cross-age tutor or adult for each group of four students to read the question cards as they are taken from the paper bags.
4. Other counters can be used instead of Unifix cubes.
5. For *Part Two*, duplicate one set of *Tool Cards* and one set of *Nursery Rhyme Cards* (pictures only) for each student. White bond is suggested for both the sets of cards because the students will glue these to construction paper.

Procedure
Part One
1. As a whole class, review the classic versions of those nursery rhymes which are represented on the *Nursery Rhyme Cards*.
2. Distribute Unifix cubes and one paper bag that will serve as a grab bag to each group of four students. Direct one student to choose a card from the bag.

45

Have the student either read the question or direct the reader assistant to read the card to the group.

3. If the student answers the question with a reasonably correct answer, he or she should take a Unifix cube and pass the bag to the next student in the group. If that student cannot answer the question, the question is passed to another student in the group and no cube is taken. Note: There may be more than one correct response to the questions.

4. Place the card back in the bag each time so that other students will have a chance to draw that same card.

5. Continue this procedure with each of the students in the group. Each time a student answers a question correctly, he or she is to take a cube.

6. The object of the game is to collect five Unifix cubes. Once all the students have collected five cubes, or more, the game ends.

Part Two
Assessment
1. Hand out one set of *Nursery Rhyme Cards* (pictures only) and one set of *Tool Cards* to each student along with the construction paper, glue, and scissors.

2. Direct the students to cut out and match the *Nursery Rhyme Cards* to the corresponding *Tool Cards* which could be used in the rhyme. Direct them to glue these paired cards onto a large piece of construction paper.

Discussion
1. Name some of the tools that were used in the nursery rhymes. Describe the job these tools were used for in the rhymes.

2. Have you ever used any of these tools? Describe how you used them.

3. Tell the class about a tool that could have been used in a nursery rhyme that is different than the one described in the rhyme.

4. Why do we use tools?

5. Which of these tools could be used in more than one nursery rhyme? Describe the job these tools would be used for in the rhymes.

Extension
Prepare additional question cards by asking the students to dictate questions about tools used in their favorite stories such as *Goldilocks and The Three Bears, The Three Little Pigs,* and others. Use these new cards in the grab bags and repeat the activity.

Curriculum Correlation
Literature
   Smith, Jessie Willcox. *The Jessie Willcox Smith Mother Goose.* Derrydale Books. NY. 1986.

Nursery Rhyme Cards

Nursery Rhyme Questions

Jack was nimble, Jack was quick, Jack jumped high to miss the wick. What was the tool that clever Jack used To keep the candle straight as a stick?	Black sheep, black sheep, Have some wool they need to keep. What's a tool that they can use To keep it safe for those they choose?	Little Boy Blue had lots to do, He called his sheep and cattle too. What was the tool he needed to use So all of his animals heard the news?
The three little kittens washed their mittens. What tools would be fine to hang them on the line?	Jack and Jill went up a hill. What was the tool they wanted to fill?	Little Miss Muffet ate curds and whey, Such wonderful manners she used today! What was the tool she used to eat A snack while keeping nice and neat?
Mary, Mary, quite contrary, How does your garden grow? What is a tool that she could use To place her plants all in a row?	Little Jack Horner sat in a corner Wearing his Christmas tie. What tool did he use to find a plum So he could eat some pie?	Hickory Dickory Dock, The mouse ran up the clock. What tool would he need To get there with speed?

Nursery Rhyme Tool Cards

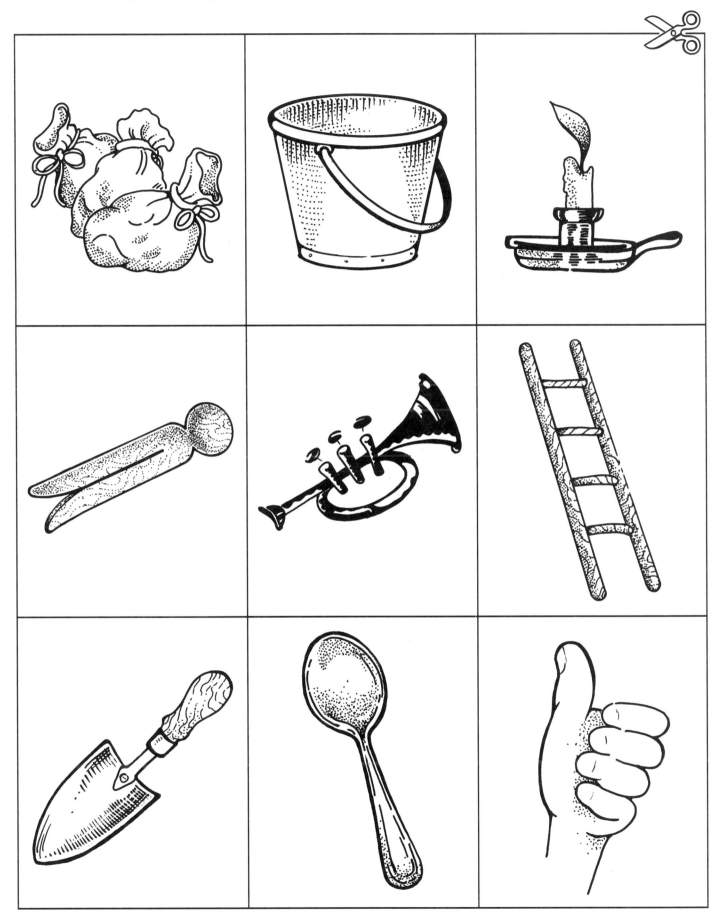

Tools Are Handy

Words by Suzy Gazlay

Tune: London Bridge

Tools are part of ev - 'ry day,

In my work, in my play;

Help - ing me in ev - 'ry way,

Tools are hand - y!

If I'd like to build a boat,
Sew a button on my coat,
Draw a picture, write a note,
Tools are handy!

If I want to cut my meat,
Pound a nail, make a treat,
Serve the food I want to eat,
Tools are handy!

If I'd like to see a star,
Measure high or wide or far,
Magnify what's in a jar, Tools are handy!

UNDER CONSTRUCTION

50

Design

Designed by Me!

Topic
Technology: Design

Key Question
What is design?

Focus
The students will explore design through a study of clothing.

Guiding Documents
Project 2061 Benchmark
* *Describing things as accurately as possible is important in science because it enables people to compare their observations with those of others.*

NRC Standards
* *Ask a question about objects in the environment.*
* *Scientists use different kinds of investigations depending on the questions they are trying to answer. Types of investigations include describing objects, events, and organisms; classifying them; and doing a fair test (experimenting).*

Science
Technology
   design

Integrated Processes
Observing
Comparing and contrasting
Communicating
Applying

Materials
For the class:
   a variety of types of clothing

For each student:
   one *Event Tag*
   one *Designed by Me!* recording book

Background Information
   Webster defines design as: "1. to make preliminary sketches of; sketch a pattern or outline for; plan 2. to plan and carry out 3. a thing planned for or outcome aimed at."

Before young learners can begin to apply Webster's definition of *design*, they need to observe and explore the design of objects in their world. They should begin to understand that the design of an object is sometimes based on the materials with which it will be used or from which it is made, as well as the purpose of the object. They will also come to understand that people change the design of objects to improve the use of the object or to adjust to the demands of the materials. Their attention will be directed to their own clothing.

Management
1. Gather a variety of clothes. Bring in samples of clothes that are designed for your current season and another set for a different season. Also include clothes for play as well as for party and for pretend (for example: uniforms, costumes, etc.)
2. If possible, locate another collection of clothes that are out of style.
3. Duplicate one *Event Tag*, one *Designed by Me!* recording book, and one paper doll for each student.

Procedure
Part One
1. Invite a student to the front of the class. Ask the class to discuss the clothes that the student is wearing. Talk about the design, whether the shirt or dress is short sleeved or long sleeved, whether it is made from a light material or heavy material, etc.
2. Repeat by observing the clothing of several other students.
3. Show the collection of clothing you have gathered (excluding the out-of-style clothes). Ask the students to sort this clothing and to discuss their rules for the sorting categories.
4. Introduce the word *design* to them through your discussion of the clothes. Discuss clothing which has the same design and samples which are of a different design. Sleeveless could be a design for warm weather where long sleeves could be a design for colder weather. Compare and contrast the differences in clothing designed for cold weather with clothes for warm weather.
5. Discuss the differences in clothes designed for play from those designed for a party of a more formal gathering.
6. Talk about the purpose of each of the pieces of clothing in the collection.
7. Lead the students to the understanding that the clothes were designed for particular purposes. These purposes may be for fashion, for weather, for comfort, for protection, etc.

8. Be cautious not to stereotype designs of clothes. Long sleeves are not necessarily only designed for warmth. Discuss with the class that there are many people who wear long-sleeved clothing in the warmest of seasons to protect their skin from the hot sun.

9. Introduce the set of clothes which are out of style. Discuss the purpose of these clothes and their design.

Part Two
1. Gather the students together and place the collection of clothing in the center of the class.

2. Distribute one *Event Tag* to each set of two students. Direct each set of partners to choose clothes from the collection that would be appropriate for the event listed.

3. Allow students to "show and tell" their choices and to explain why they chose these items. Encourage the groups to describe the *design* of these clothes.

4. Collect the *Event Tags*.

Part Three
1. Redistribute the *Event Tags*, being careful to give each student a different tag than they used in *Part 2*.

2. Hand out one *Designed by Me!* book and paper doll to each student.

3. Instruct the students to glue the *Event Tag* to the appropriate space on the *Designed by Me!* book. Direct them to design and illustrate appropriate clothing for the paper doll to use for the event they received. Tell the students to glue these clothes onto the paper doll. Explain how to fold and cut out the pop-up space inside the book. Add the front cover and attach the paper doll. (See illustration.)

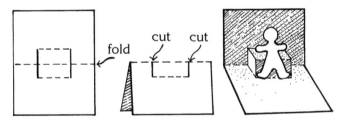

4. Allow time for the students to share their designs with the class.

Discussion
1. What is a design?
2. Compare and contrast the design of two different pieces of clothing.
3. Where do you think these clothes should be worn? [to a party, to school, camping, costume party]
4. Why do people change the design of clothes?
5. Who do you think designs clothes?
6. Does the material used to make the clothes change the design? Explain.
7. Which kind of clothes do you like to wear? Explain.

Extensions
1. Invite a fashion designer and/or tailor to the classroom to share experiences from their work.
2. After the students finish illustrating their designs in *Part 3*, provide material for them to actually make the clothing.
3. Bring in books to share different fashions through history and discuss the changes and reasons for these changes through the years.
4. Bring in books or samples of clothing to share fashions from different cultures.
5. Have students analyze a dress or shirt pattern to see how the pieces go together for making a garment. Pattern books are wonderful resources for showing various designs. Ask at your local fabric stores to see if they will let you have expired books to use with your students.

Designed by Me!

Event Tags

You are going to a costume party.	You are going to a birthday party.	You are going to the beach to swim.
You are going to play in the snow.	You are going to a wedding.	You are going to bed.
You are going to school.	You are going to rake leaves.	You are going for a walk in the rain.
You are going camping.	You are going to be a doctor.	You are going to be a cowboy.

 # Designed by Me!

Paper Dolls

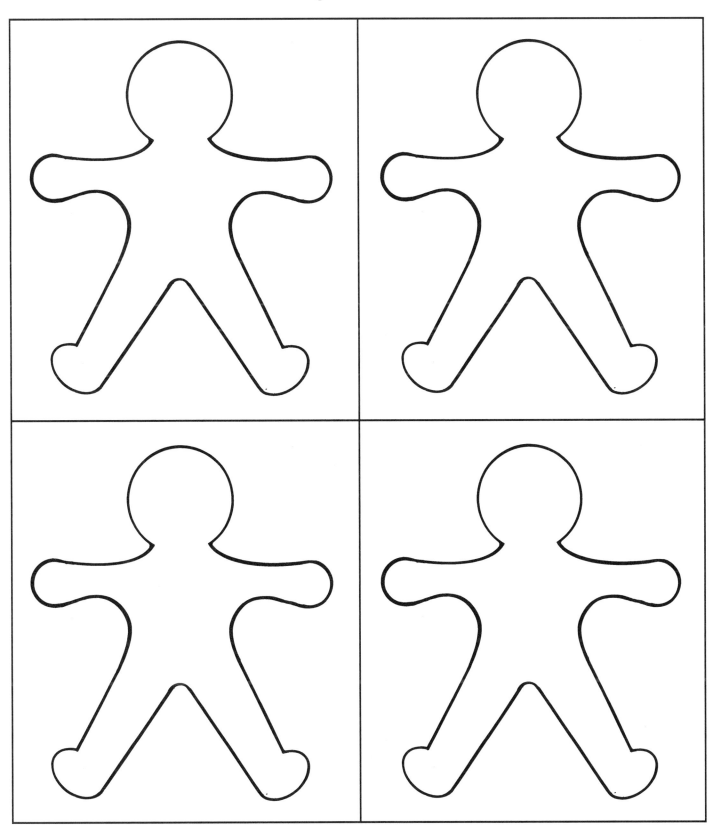

Designed by Me!

Name_____

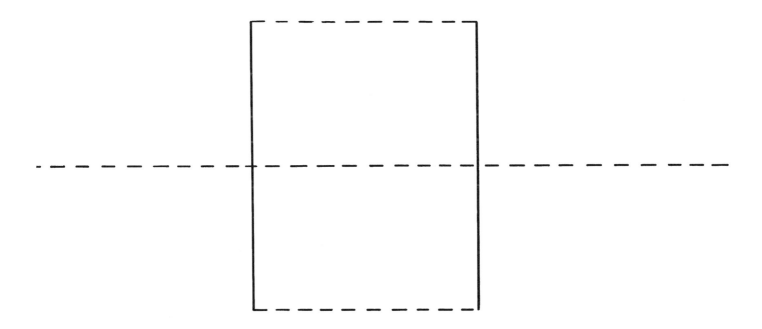

57

CAN It Open?

Topic
Technology: Design

Key Question
Why do designs change?

Focus
The students will explore different designs of can openers.

Guiding Documents
Project 2061 Benchmarks
- *People can use objects and ways of doing things to solve problems.*
- *Tools are used to do things better or more easily and to do some things that could not otherwise be done at all.*
- *People, alone or in groups, are always inventing new ways to solve problems and get work done. The tools and ways of doing things that people have invented affect all aspects of life.*

NRC Standards
- *People have always had problems and invented tools and techniques (ways of doing something) to solve problems.*
- *Some objects occur in nature; others have been designed and made by people to solve human problems and enhance the quality of life.*
- *Objects can be categorized into two groups, natural and designed.*

Science
Technology
    design

Integrated Processes
Observing
Contrasting and comparing
Communicating
Applying

Materials
For the class:
    three or four kinds of can openers (see *Management 1*)
    several large cans (see *Management 2*)

Background Information
The focus of this study of design is to help the students begin to construct an understanding that objects, including tools, are designed with a purpose in mind. The design of a tool is sometimes based on the materials used, as well as the project or job to be done with the use of this tool.

The students will discover that designs are changed to improve the use of the object or to adjust to the demands of the materials or the job. In this activity, they will look at variations of a common, everyday tool — the can opener. They will be asked to think about why tools were designed as they are.

Management
1. Gather a variety of can openers. Try to include a sample of each of the types pictured.

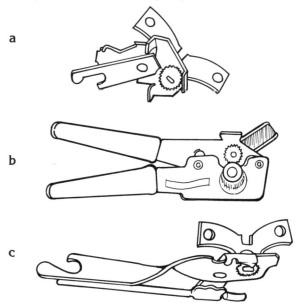

a

b

c

Figure 1

2. Collect several large, empty, metal cans which are open on one end only. Some suggestions are: coffee cans from the teacher's lounge, #10 cans from the cafeteria, etc. Using masking tape, cover any rough or sharp edges.

Procedure
1. After the students have experienced the *Project Cards* and *Challenge Cards* in *Materials Matter* and

Tools of the Trade, ask them to think about the tools they used with these projects. Discuss the job these tools helped the student to do and how the tool is designed to help with this job.

2. Ask the students if they can think of a way to change the tool that would have made the job even easier.
3. Show the students the can opener such as illustration *a* in *Figure 1*. Pass around the opener and a can, giving each student a chance to turn the opener. (Use additional cans if the students do succeed in opening the can before the entire class has a chance to turn the opener.)
4. Now show them the can opener with handles (illustration *b*). Ask them to try opening another can using this opener.
5. Repeat this procedure with all the can openers.
6. Discuss with them the difference in the design between the can openers.

Discussion
1. What is a design?
2. Compare the design of two or more different can openers. Which do you think is best? Why?
3. What is the purpose of a can opener? Why do you think they are designed this way? Can you think of a different way to design a can opener? Describe it.
4. Does a can opener have more than one part? Explain.
5. What do you think would happen if we changed the design and took away some of the parts? Discuss with the students how some things have more than one part and that the tool usually needs all the parts working together to work efficiently.
6. Why do people change the design of things?
7. Who do you think designs objects, tools, etc.? Is a can opener made and designed by a person or by nature? Explain how you know this.

Extension
Bring in other items such as a collection of ink pens. Explore with the children the evolution of the design of the pens from the quill to the fountain pen to the ball point pen. Discuss the changes in the design and possible reasons for these changes.

Topic
Technology: Design

Key Questions
How can you make a container by cutting and folding paper?
How does the design affect the use of the container?

Focus
The students will discover how a variety of containers are designed.

Guiding Documents
Project 2061 Benchmarks
- *Circles, squares, triangles, and other shapes can be found in things in nature and in things that people build.*
- *When trying to build something or get something to work better, it usually helps to follow directions if there are any or to ask someone who has done it before for suggestions.*
- *Several steps are usually involved in making things.*

NCTM Standards
- *Develop spatial sense*
- *Recognize and appreciate geometry in their world*
- *Develop the process of measuring and concepts related to units of measurement*
- *Make and use measurements in problems and everyday situations*

Math
Geometry
Measurement
Equalities and inequalities

Science
Technology
design

Integrated Processes
Observing
Recording
Comparing and contrasting
Communicating
Applying

Materials
For each student
recycled greeting cards (see *Management 1*)
patterns for containers
scissors
transparent tape or glue
recycled box containers from home (see *Management 4*)
Teddy Bear Counters or Unifix cubes (see *Management 5*)

For the class:
two sets of boxes (see *Management 6*)
a variety of art materials (see *Management 3*)

Background Information
Have you ever wondered how the variety of containers we use every day are designed and constructed? This activity will show the students the process of building boxes and other containers with which they may be acquainted. The students will discover that the finished product looks much different than its original, flattened pattern. Through folding, cutting, and sometimes gluing or taping, the students will construct a variety of containers. Although step-by-step instructions are included, it is suggested that the teacher takes the students through the steps for one or two containers, and then allows them time to figure out how to build a given model and devise their own containers.

Through the process of building these containers, the students should gain a greater awareness and interest in how things are designed and constructed. To heighten student interest, allow them to bring containers from home which can be taken apart, examined, and discussed by the class.

Management
1. Copy the container templates onto lightweight cardstock. If lightweight cardstock is not available, paper will also work. *Box 1* does not need a template; it is made by having students follow oral directions. Old greeting cards brought from home are used for *Box 1*. For ease of construction, the greeting card should be no smaller that 14 cm x 11 cm (5 1/2" x 4 1/4"). The front of the card forms the lid of the box, while the back of the card forms the bottom. If used greeting cards are not available, use cardstock.
2. Plan to teach the construction of only one container each day. Allow ample time to make and

remake several similar containers so they can practice their skills.

3. Gather crayons, paints, stamps, stickers, or other art materials so students can decorate their containers. These containers make great packages for gifts for the family for various occasions. Once decorated, the container itself can be the gift!

4. Ask the students to bring in two identical boxes from home. Encourage them to bring in containers such as cereal boxes, toothpaste boxes, etc.

5. Teddy Bear Counters, Unifix cubes, or other such manipulatives can be used as non-customary units to measure the volume of the containers.

6. Collect two sets of two identical boxes. Make sure the sets are of similar size but of different shapes. Take apart and flatten one box from each set. Do not cut off any flaps, etc.

Procedure

1. Show the class the two boxes which are still intact along with their "twins" that you have carefully disassembled. Ask students to predict which intact box matches each disassembled box.

2. To help students understand the question, reconstruct each of the boxes in front of the class. Discuss how the boxes are formed and are designed.

3. Tell the class that they are going to make several different boxes and containers to discover how they are designed and constructed.

4. Distribute one recycled greeting card to each student.

5. Use the *Box 1 Instructions* to take the students through the process of folding and cutting the greeting card to make the box and lid.

6. On different days, make other containers using the instructions and patterns provided.

7. Discuss what items would fit into their containers. Be sure to include vocabulary such as *too big, too small,* and *just right.*

8. Discuss how the design of each container changes its use.

9. Discuss the different shapes formed from the process of folding and cutting as they make their containers.

10. Using Teddy Bear Counters, or other manipulatives, direct the students to compare the capacities of the different containers.

11. Direct them to look at the boxes they brought from home. Ask them to carefully take one of the boxes apart in order to display the form of the unconstructed box. Ask the students to study the form and to try to put the box back together again. If necessary, allow them to use the companion box as a model.

Discussion

1. What shapes did you make when you were folding your container?

2. Describe how you made your container.

3. Show something that would fit *just right* into your container.

4. Show something that would be *too big* for your container.

5. Which of your containers held the most Teddy Bear Counters?

6. How do you think the shape or design of the container affects the amount it holds?

7. How did the design of each container determine how its used?

8. Tell how one container is like another. Tell how it is different.

9. How do you think companies make boxes?

10. Describe how the box you brought from home is constructed.

Extensions

1. Use the bulletin board to display the opened, disassembled forms of different containers, placed so the interiors of the containers are visible to the students. On the same board, but in a different arrangement, display a completed container for each. Ask the students to try to match the form to the completed container.

2. Ask the students to design a new container. Direct them to make a template and to then duplicate the template and to construct the container. Display each design.

3. Have students look at the forms of envelopes, shopping bags, and grocery sacks.

4. If you have a container manufacturer in your area, take the students on a field trip to watch the machines that fold, glue, and form containers.

5. Use these containers as packaging for a gift for the family for the holidays.

6. Use the activity *Made By Nature and Made By Me!* to make a gift that could fit into the containers.

Fold to Hold

To make this box and lid, use an old greeting card which is cut along the fold to make two same-size pieces.

Box 1 Instructions

Fold as indicated by the broken lines.

1. Fold one piece of the card in half and crease.

2. Fold both ends in to the crease you just made. Crease these new folds.

3. Open the card. You should see four sections

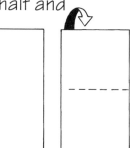

4. Fold the card in half the other way and crease.

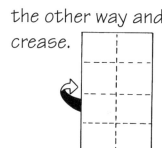

5. Fold both sides in to the crease you just made. Crease these new folds.

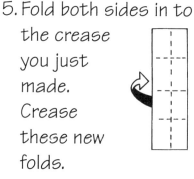

6. Open the card. You should see 16 sections.

7. Make four cuts along the folds as indicated.

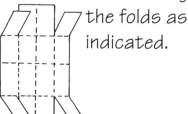

8. Fold the corner pieces in on both ends.

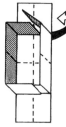

9. Fold the end flaps up and over both ends.

10. Do the same with the second half of the card.

Fold to Hold

1. Copy on paper or lightweight cardstock.
2. Cut along solid lines.
3. Fold along broken lines.
4. Glue tabs.

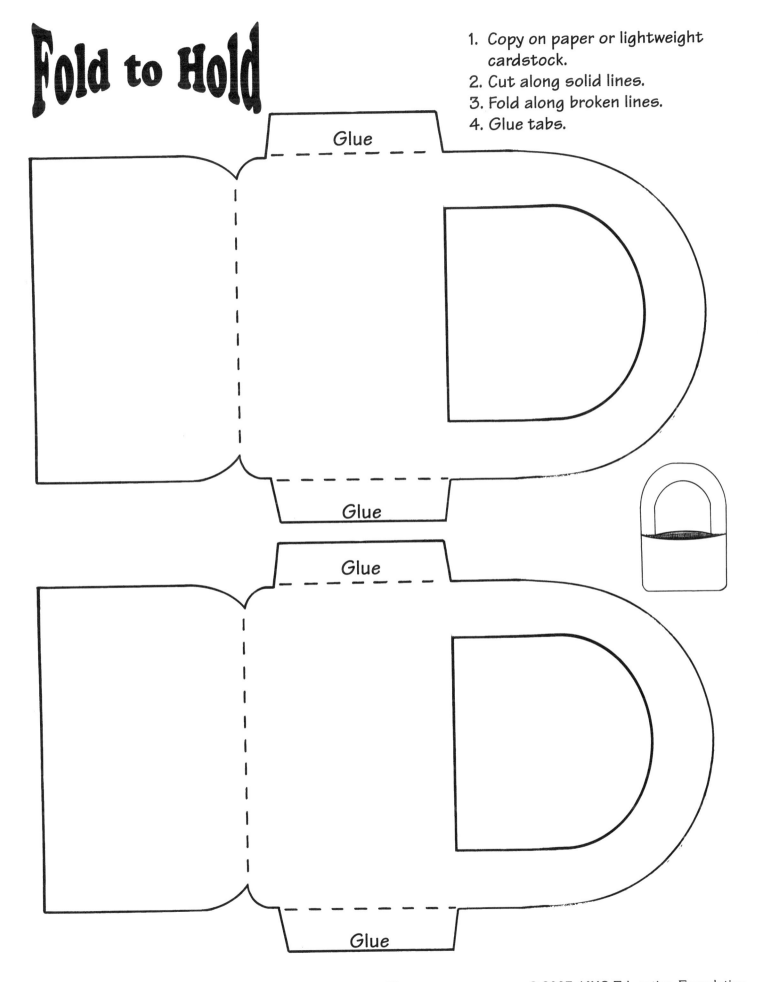

Glue

Glue

Glue

Glue

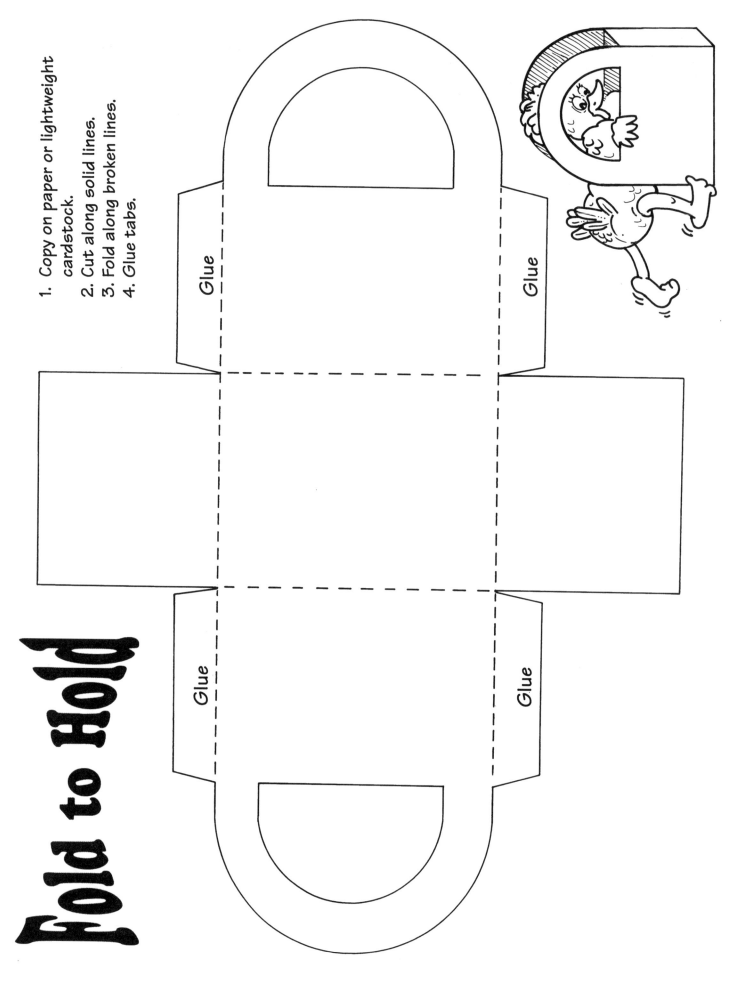

Fold to Hold

1. Copy on paper or lightweight cardstock.
2. Cut along solid lines.
3. Fold along broken lines.
4. Glue tabs.

Glue

Glue

Glue

Glue

Fold to Hold

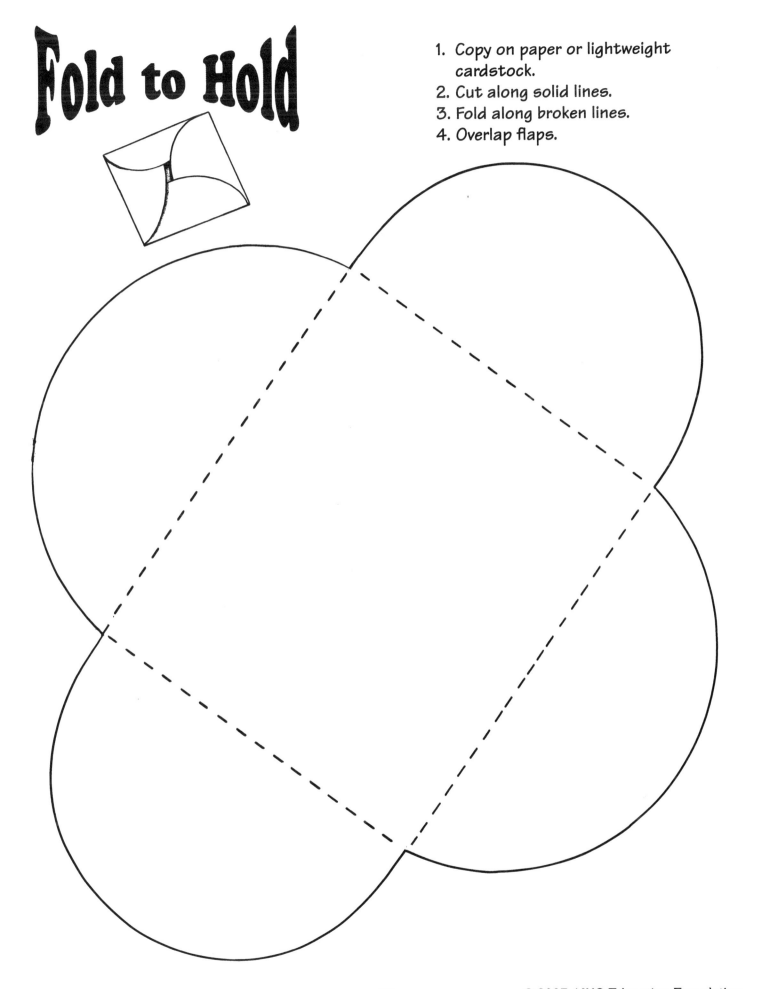

1. Copy on paper or lightweight cardstock.
2. Cut along solid lines.
3. Fold along broken lines.
4. Overlap flaps.

UNDER CONSTRUCTION
65
© 2007 AIMS Education Foundation

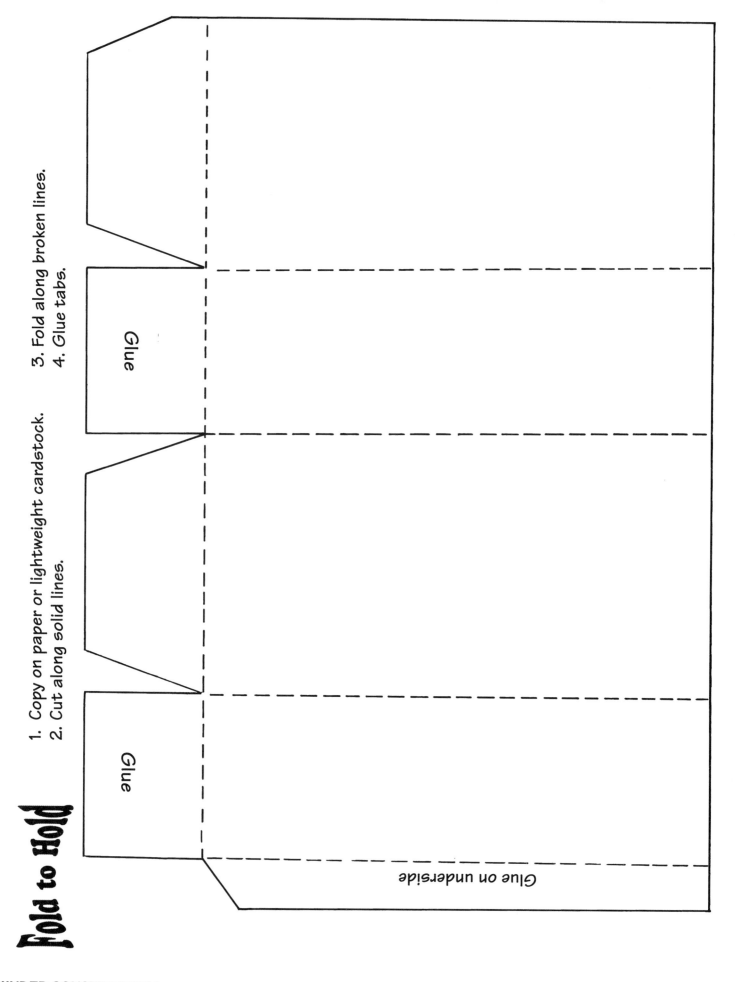

Fold to Hold

1. Copy on paper or lightweight cardstock.
2. Cut along solid lines.

3. Fold along broken lines.
4. Glue tabs.

Glue

Glue

Glue on underside

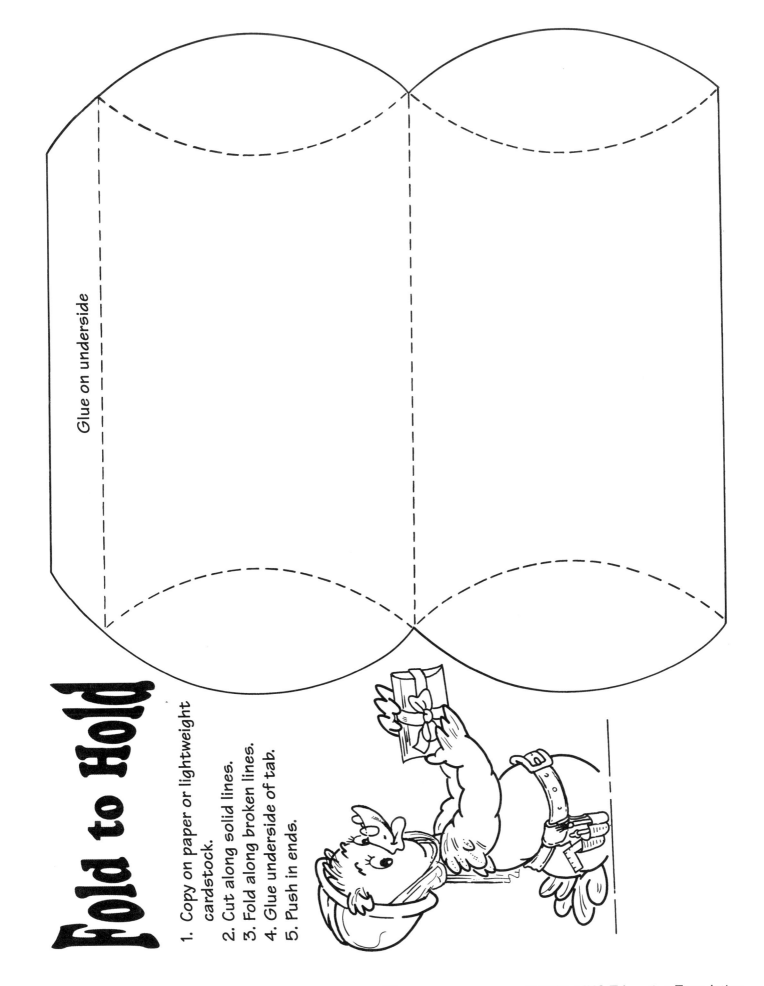

Fold to Hold

1. Copy on paper or lightweight cardstock.
2. Cut along solid lines.
3. Fold along broken lines.
4. Glue underside of tab.
5. Push in ends.

Glue on underside

Bag It

Topic
Technology: Materials/Design

Key Question
What material and design is best for a bag?

Focus
The students will design a bag for a particular purpose.

Guiding Documents
Project 2061 Benchmark
- *Some kinds of materials are better than others for making any particular thing. Materials that are better in some ways (such as stronger or cheaper) may be worse in other ways (heavier or harder to cut).*

NRC Standards
- *People have always had problems and invented tools and techniques (ways of doing something) to solve problems.*
- *Abilities of Technological Design*
  - *Identify a simple problem.*
  - *Propose a solution.*
  - *Implement proposed solutions.*
  - *Evaluate a product or design.*
  - *Communicate a problem, design, and solution.*

NCTM Standard
- *Collect, organize and describe data*

Math
Counting
Equalities and inequalities

Science
Technology
    characteristics of materials and design

Integrated Processes
Observing
Comparing and contrasting
Recording data
Communicating
Applying

Materials
For the class:
    a variety of bags made from cloth, paper, and plastic (see *Management 1*)
    two large floor graphs and labels (see *Management 2*)
    a variety of materials to construct bags (see *Management 3*)
    glue
    scissors
    transparent tape

For each student:
    Family Note (see *Management 1*)
    one set *Bag It-File Folders Project* pages
    envelope
    2 paper fasteners
    jumbo paper clip

Background Information
Through their observations, students will begin to construct an understanding of the suitability of different materials and designs of bags for serving a variety of purposes. They will apply their knowledge of materials and designs to the making of a bag for a specific purpose.

Management
1. To generate a collection of various types of bags, duplicate and send home one *Family Note* for each student.
2. Prepare two large floor graphs and blank labels. The students will generate and record the information for the labels in *Parts One* and *Two* of the activity.
3. For *Part Three*, provide a variety of materials such as: plastic from garbage bags, food wrap, brown mailing paper, butcher paper, newspaper, and scrap paper. CAUTION: When using plastic, be sure to guard against students placing the plastic over their nose or mouth.
4. Duplicate one set *Bag It-File Folders Project** pages per student. For young learners, you may want to prepare the envelopes beforehand; then the students can cut and fold the folders.

Procedure

Part One

1. Discuss the variety of bags the students brought in. Generate categories for a class floor graph to display the different bags. Direct the students to make category labels and to place them on the floor graph. Category examples might include: different materials such as paper, plastic, and cloth **or** different uses such as party, lunch, grocery, garbage, and sandwich.
2. Direct the students to place their bags in the appropriate spaces on the graph.
3. Ask the students to name possible uses for the different types of bags.
4. Discuss how these are all bags, but they are made with different materials or have different designs.

Part Two

1. Tell the students to gather all the *large* bags.
2. Challenge them to design a plan to test these bags for strength. If they have a difficult time designing an experiment, suggest using a measured amount of sand, cans of food, etc., in each bag.
3. Distribute a different type of large bag to each group of four students. Direct them to perform the strength test as agreed upon by the class.
4. Use a floor graph to record the class results. Have the students make graph labels. The title of the graph could be "How Strong Is My Bag?" The column labels might be: Held 2 cans of food, 3 cans of food, 4 cans, etc. Another suggestion is "Which bag was the strongest?" The column labels could be, plastic, paper, cloth, etc.
5. Tell them to place their bag in the appropriate column on the graph to record the results of their experiment.
6. Discuss the results with the class.

Part Three

1. Using the class collection of bags, direct the students to observe the design of the bags. Tell them to compare and contrast how the bags are put together, the shapes, the materials used to make the bags, etc.
2. Challenge the students to make a bag of their own. Tell them that they will need to identify a *purpose* for the bag, the *materials* they will need to make the bag, and a *design* for the bag.
3. Provide a variety of materials for the students to use. Give them time to plan, make, test, and if needed, remake their bags.
4. Distribute the *Bag It-File Folders Project* pages to each student. Assist them with the directions for the construction of the file folders and holder.
5. Tell them to use this file folder packet to record the *purpose* of their bags, the *materials* they used, and an illustration of their final *design*. Encourage the students to display their projects and to discuss the process of making the bags, how they chose the materials, and their designs.

Discussion

1. Explain the similarities and differences of the bags on our graph.
2. Choose a bag and describe what the bag would be best used to do.
3. Describe something that probably would not be a good use for this bag. Explain why.
4. Show the class a bag that would be good to use in water. Explain why.
5. Choose three bags with different closures. Explain why one would be better than another for a certain purpose.
6. Do all bags need to be strong enough to hold heavy objects? Explain. [No, it depends on the purpose or intended use of the bag. Some bags may be used to hold small things like a greeting card.]
7. Were you able to construct a bag according to what you wanted as a design? Did you change your design from what you originally thought about? Why? Describe what you could use your bag for?
8. Find another bag that could also be used for your purpose. Find a bag that would not be appropriate to use for this same purpose. Explain why you chose each one.
9. How do the materials used to make the bags make a difference in the possible uses of the bags?

Extensions

1. Suggest to the students that they design a test to determine the durability of the different materials used to make the bags. Allow them to plan and test their ideas. If they need additional guidance, suggest that they place the same item in each bag, place each bag in water for one minute. Remove and observe the results.
2. Explore the materials used and the design for other types of containers: bottles, boxes, etc.
3. Ask the students to think of something for which they would need a bag, a box, or a bottle. Challenge them to find the ideal container for their purpose and to explain why this choice is best.

* The design of the file folders was taken from *Big Book of Books and Activities* by Dinah Zike. This book is an excellent teacher resource. It includes many unique ways for students to display their work. The book can be ordered from Dinah-Might Activities, Inc., P.O. Box 39657, San Antonio, TX 78218 or by calling (512) 657-5951.

Bag It
Family Note

Dear Family,
We are continuing our study of technology and are now learning about design. We need a large variety of bags of different designs and sizes. We need bags made from:

plastic
cloth
paper
nylon

Thank you for your help with this collection. We will use these items to explore and discover many things in our study of technology.

Bag It
Family Note

Dear Family,
We are continuing our study of technology and are now learning about design. We need a large variety of bags of different designs and sizes. We need bags made from:

plastic
cloth
paper
nylon

Thank you for your help with this collection. We will use items to explore and discover many things in our study of technology.

Bag It— File Folders Project

Cut on solid lines.
Fold on broken lines.

Partially open a paper clip to form a clasp.

Use a paper punch to make a hole in the center of the flap. Insert a paper fastener. Place another paper fastener in the body of the envelope in a location that allows the flap to be closed without covering the second paper fastener.

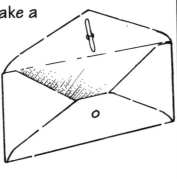

Cut out and fold the three file folders.

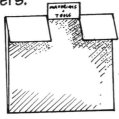

Place the file folders in the envelope. Attach the paper clip clasp to close the envelope.

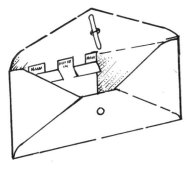

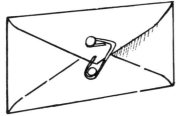

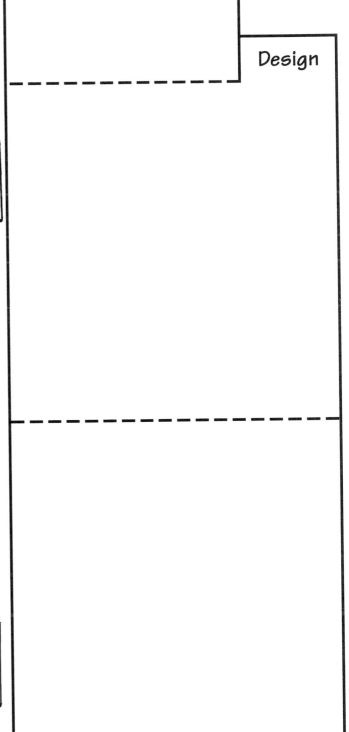

Design

Bag It - File Folders Project

Cut on solid lines.

Fold on broken lines.

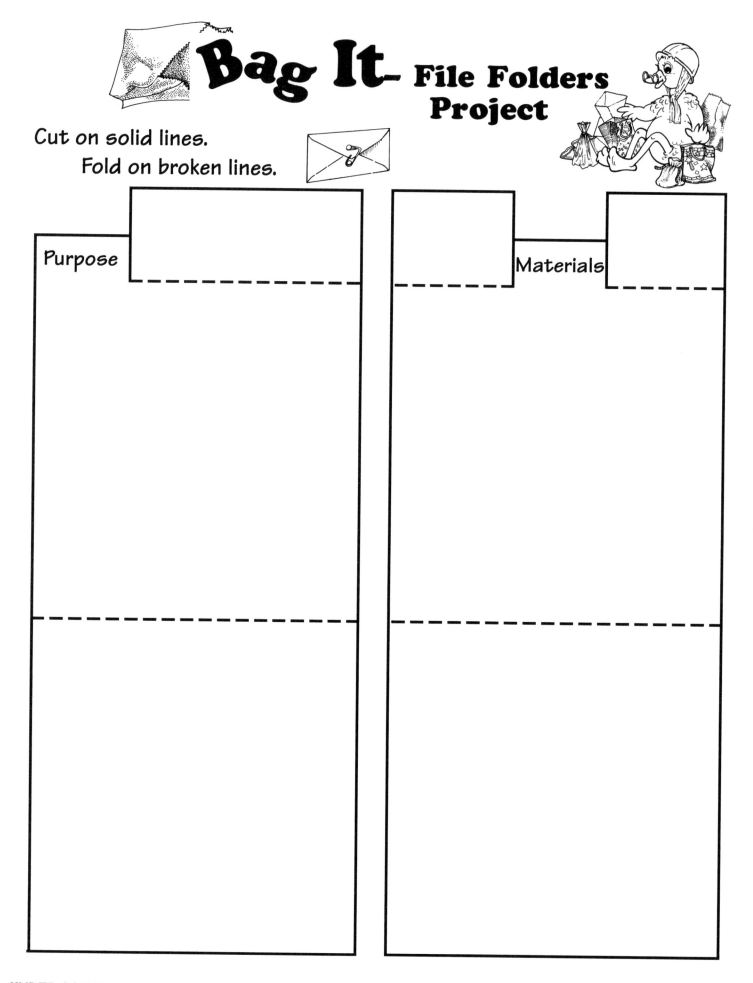

Purpose

Materials

Exploring Bridges

Topic
Technology: Design

Key Questions
What do you need to do to build a bridge?

Focus
The students will explore bridges through their block play and begin to construct their own knowledge about bridges.

Guiding Documents

Project 2061 Benchmarks
- *People may not be able to actually make or do everything that they can design.*
- *When parts are put together, they can do things that they couldn't do by themselves.*

NRC Standards
- *Scientists use different kinds of investigations depending on the questions they are trying to answer. Types of investigations include describing objects, events, and organisms; classifying them; and doing a fair test (experimenting).*
- *Abilities of Technological Design:*
  - *Identify a simple problem.*
  - *Propose a solution.*
  - *Implement proposed solutions.*
  - *Evaluate a product or design.*
  - *Communicate a problem, design, and solution.*

NCTM Standards
- *Develop and apply strategies to solve a wide variety of problems*
- *Apply estimation in working with quantities, measurement, computation, and problem solving*

Math
Measurement
Estimation

Science
Technology
   design

Integrated Processes
Observing
Comparing and contrasting
Communicating
Applying

Materials
For the class:
   a large supply of building materials (see *Management 1*)
   a supply of connecting materials (see *Management 2*)
   materials for constructing the following: (see *Management 3*)
   cities
   a river
   a ravine
   one set of the *Study Prints* (see *Management 5*)

Background Information
Primary students are natural builders. This can readily be seen during block play when they build things in different ways and for different purposes.

Before any formal instruction about bridges is begun, allow the students to explore different materials for building them. This activity is designed to allow the students to construct their own knowledge about bridges through a trial-and-error experience using a variety of building materials.

Management
1. Provide milk cartons, large and small wooden blocks, scrap lumber, cardboard boxes, paper towel tubes, foam rubber blocks — anything you can think of that might be appropriate for building a classroom bridge.
2. Supply things that can be used to connect the building materials such as masking tape, glue, string, etc.
3. Provide large paper for the children to make a river on the floor of your classroom. Cardboard boxes and other classroom manipulative materials will be good for building cities, a ravine, etc.
4. Be careful not to give direct answers. Let students discover the strengths and weaknesses of the different materials.
5. Duplicate the *Study Prints*, color, and laminate for continued use.

Procedure
1. Read the story *The Three Billy Goats Gruff* to your students. You may want to read several versions which are listed in the *Bibliography*.

2. Discuss the bridges in the story and the different designs and materials used to build the bridges.
3. Ask why the goats needed to use a bridge.
4. Encourage the students to think about bridges in their area or bridges they have seen in books, movies, etc.
5. Ask them to explain why the bridges are there and what they are used for. (Try to lead the students into a discussion that bridges are used to get from one place to another that may be difficult to get to without the bridge.)
6. Show the *Study Prints*. Discuss what the bridges are crossing and possible reasons for the bridges.
7. While the students are building bridges in a free exploration center, help them find materials to construct a paper river, cardboard buildings for a city, a ravine, etc. for their bridges to span.
8. As the children build, informally circulate through the classroom asking the following questions.

Discussion

1. Why are bridges used?
2. What makes a bridge strong?
3. Name some materials that will not make a strong bridge. Why do they seem weak?

4. Name some building materials that can be used to build a strong bridge. Why do they seem strong?
5. What did you use to build your bridge?
6. What is the strongest part of your bridge?
7. What is the weakest part of your bridge?
8. How could you make your bridge stronger?
9. If you took away one part of your bridge, would it still work as a bridge? How about two parts? Explain.
10. How are the different parts of your bridge important to the whole bridge?
11. What do you think bridges that people walk on, drive on, etc. are made from?
12. How are our bridges similar to those people use to drive over? How are they different?

Extensions

1. Build a river in the sand box and have the students design a bridge to span this river.
2. Build a city out of LEGO® elements and other blocks. Challenge the students to build a bridge over the city. Caution them not to place their supports in the middle of a street and not to block traffic!
3. Show study prints of different designs of bridges and encourage the students to build models of these designs.

75

76

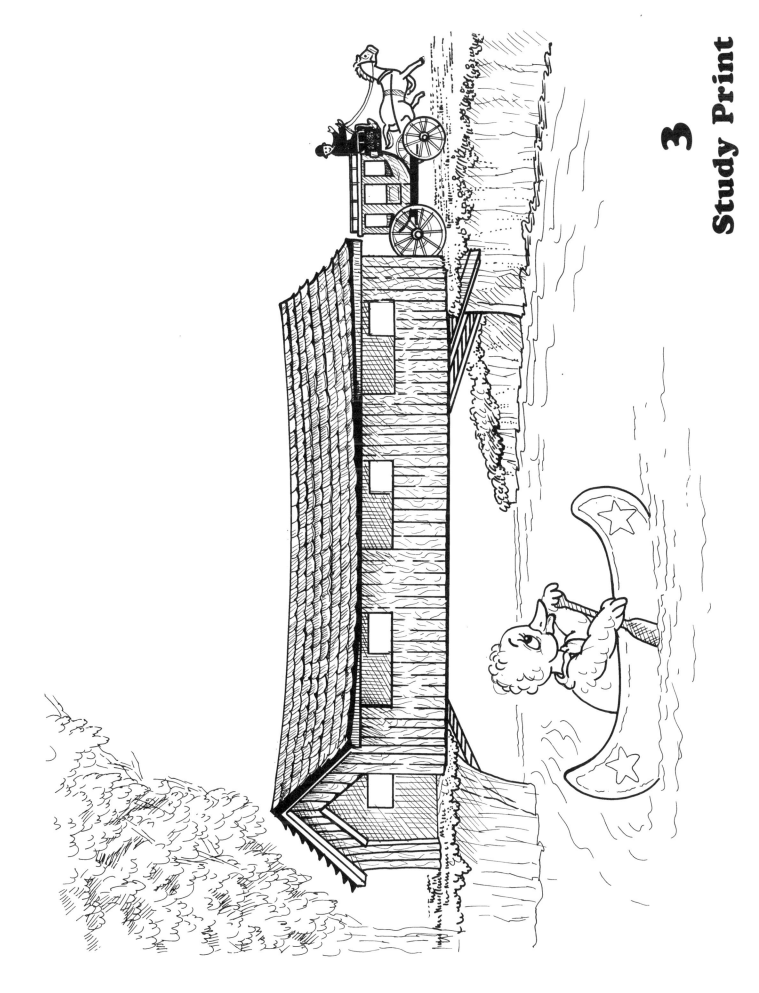

UNDER CONSTRUCTION

77

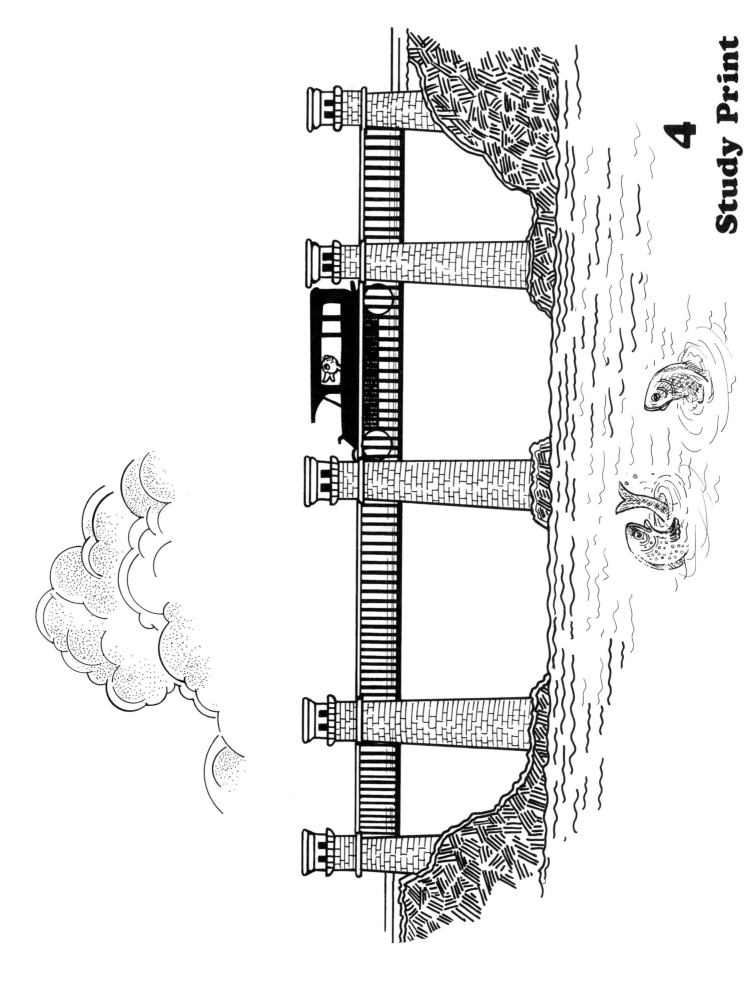

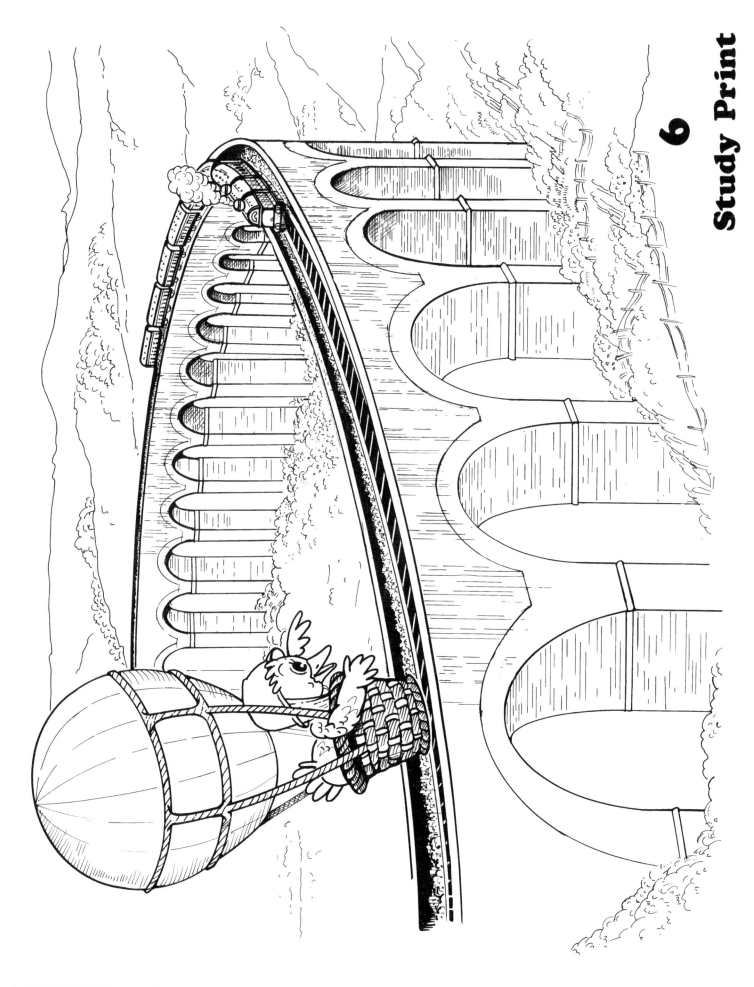

UNDER CONSTRUCTION

80

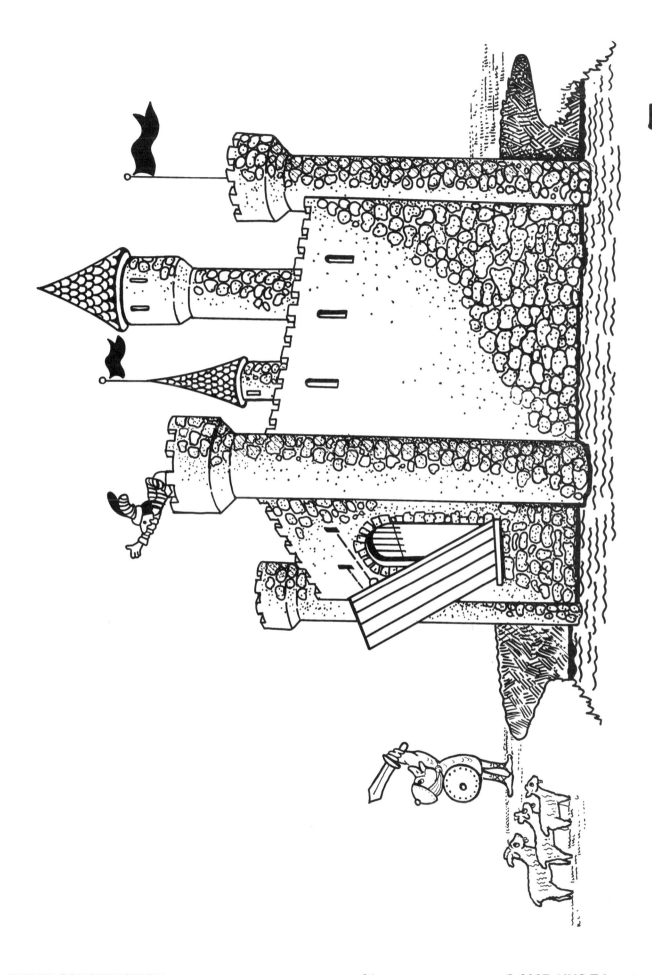

81

TRAFFIC PATROL

Hold The Load

Topic
Technology: Design

Key Question
How does the design of a paper bridge affect how much it will support?

Focus
The students will build and then compare and contrast the differences in the strength between two designs of bridges: a beam and an arch bridge.

Guiding Documents
Project 2061 Benchmarks
- *People may not be able to actually make or do everything that they can design.*
- *When parts are put together, they can do things that they couldn't do by themselves.*

NRC Standard
- *Scientists use different kinds of investigations depending on the questions they are trying to answer. Types of investigations include describing objects, events, and organisms; classifying them; and doing a fair test.*

NCTM Standards
- *Develop the process of measuring and concepts related to units of measurement*
- *Make and use measurements in problem and everyday situations*

Math
Measurement
  height

Science
Technology
  design

Integrated Processes
Observing
Recording
Comparing and contrasting
Communicating
Applying

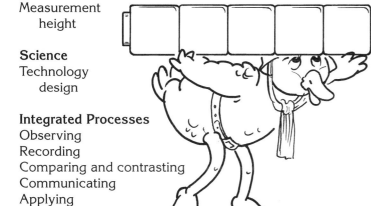

Materials
For each pair of students in a small group:
  14 Unifix cubes
  a minimum of 10 pennies
  cardstock bridge (see *Management 2*)
  blocks (see *Management 3*)
  one *Hold the Load* recording page
  a watercolor marker (see *Management 4*)
  paper towel or a soft cloth (see *Management 4*)
  masking tape

For the class:
  Study Prints from *Exploring Bridges*

Background Information
This activity allows students to explore the differences between the strengths of a beam-style bridge versus an arch-style bridge using like materials in the structures. It is intended for the students to simply realize that there are different bridge designs and to experience the difference in the strengths between these two structural designs. The following information regarding force and strength is for the teacher and not intended for the students.

It is easier to break an egg squeezing it across the sides than by squeezing it from end to end. The reason is not that the eggshell is thicker at the ends, but because of the difference in the shell's curvature at both ends. **Some shapes are stronger than others** — a fact that is important in the design of every structure, from subways to sports arenas, to bridges, and even to eggs.

This activity will deal with a simple beam bridge and an arch bridge. Some of the earliest bridges made by people were logs laid between opposite banks of a ditch or stream (see *Study Print 1*). The log, used in this way, is an example of a beam-style bridge. Beam bridges may be made of wood, stone, concrete, iron, or steel. In many cases they are a very efficient design; however, the beams will bend or break if too much weight is put on them. The load on a beam is usually greater at certain points than others, so that the structure has weak spots where it will tend to give way. One way to avoid this problem is to make the beam thicker or to build it using a very strong material. Another alternative is to cap both ends and make an I-beam. *Study Prints 2, 3, 4, 7, and 8* are all examples of the beam-type structure of bridges. They show a variety of materials, designs, and uses for the beam bridge. The students will have most likely built this type of bridge in the activity *Exploring Bridges*.

A different type of structure, such as an arch bridge, may be used instead of the beam. The shape of an arch gives the materials more strength. The downward force of the load is carried away from the middle of the arch, along the arch's smoothly curving sides to its supports. This spreads out the effect of any load. Examples of the arch bridge can be found in *Study Prints 5 and 6*.

Management

1. It is best to teach this lesson in a small group of six to eight students. You will need to plan your materials for three to four sets of partners for each group of students.
2. For each pair of students, cut two strips of cardstock, one that is 3″ x 11″ strip and another 3″ x 8 1/4″. On the longest strip, draw a line 1 3/4″ from each end.

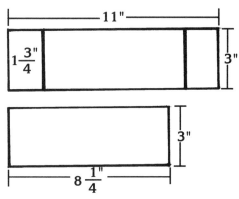

Figure 1

3. Locate four blocks 2″ x 4″ x 6″ for each set of partners. (Books can also be used instead of blocks and should be stacked to a height of 2″.)
4. Watercolor markers should be used so that the marks can be easily removed from the Unifix cubes by using a damp paper towel when the lesson is finished.
5. Duplicate one *Hold the Load* recording page per student.

Procedure

Part One

1. In a small group of students, show the *Study Prints*. Discuss the similarities and differences in these bridges.
2. Tell the class that they are going to investigate two different bridge designs to find out which one is the strongest.
3. Draw attention to *Study Prints 1*. Inform the students that they are going to build a model of such a bridge using a strip of cardstock.
4. Hand out the strips of cardstock, blocks, and Unifix cubes to each set of partners.
5. Assist the students in building their bridges by directing them to place two blocks on a table 10 Unifix cubes apart.
6. Tell them to lay their bridge, the longest strip, across the gap between the blocks. Direct their attention to the lines on the bridge. Tell them to align these lines with the inside edges of their blocks. Instruct them to place a second block on top of each of the ends of the strip.

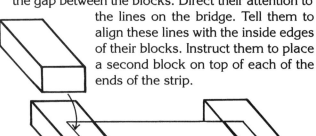

Figure 2

7. Using Unifix cubes, direct the students to measure the height of the center of the bridge above the table. Tell them to use a marker to make a line on the Unifix cubes to record the height. (The center of the bridge span should be approximately two Unifix cubes high.)

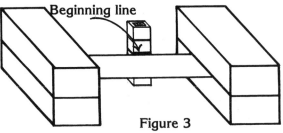

Figure 3

8. Distribute one *Hold the Load* recording page to each student. Direct the students to place the Unifix cubes next to the drawing of the cubes on the page. Tell them to draw a line on their paper in the same position as the line on their cubes. Instruct them to color in the Unifix cubes below the line they drew to represent the *Starting Measurement* for the beam bridge.
9. Tell the students to return their Unifix cubes to the center of their bridge as in *Figure 3*. Remind them to check to make sure the bridge and the marked line are even.
10. Direct the students to place a penny on the midpoint of the bridge span. Tell them to observe the line they drew on the Unifix cube and the position of their bridge. Discuss what happened.
11. Tell them to place another penny on top of the first and again observe the line they drew on the Unifix cube.
12. Continue this procedure, adding pennies to the center of the bridge span and observing the marked line and the position of the bridge until they have placed 10 pennies. Direct them to mark their Unifix cubes once more with a line to show the new height of their bridge.

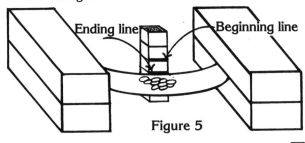

Figure 5

13. Tell the students to record their results on the *Hold the Load* recording page to show the bridge height with the added weight of the pennies.
14. Ask the students what they observed.
15. Allow them to try different ways of placing their pennies on the bridge and to compare their results.

Part Two
1. In a small group, ask partners to look at their recording pages and to recall the results of the previous activity using a beam bridge. Tell them that they will now build a model of a different shaped bridge, called an arched bridge. Review the *Study Prints* pointing out the differences in shape of the beam bridges and arch bridges. Remind them that they will be testing this bridge using the same procedure as before.
2. Distribute two strips of cardstock (3" x 11" and 3" x 8 1/4") to each group.
3. Assist the partners in building these arched bridges by instructing them to place the short strip, forming an arch, between the two bottom blocks as illustrated.

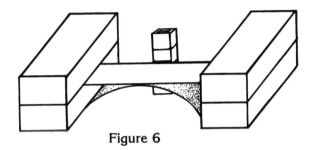

Figure 6

4. Direct them to now place the longer strip on top of the arch in the same position as they used for the beam bridge. (Remind them to align the bold lines with the inside edges of the blocks.)
5. Repeat the same procedure of marking the Unifix cubes and then adding pennies. (See *Part One* steps 7 through 15.)
6. Discuss what they discovered about the differences in strength between the two shapes of bridges.

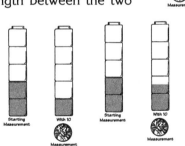

Discussion
1. What happened when you placed pennies on the beam bridge? [The bridge fell below the line.]
2. Describe how you put your pennies on the bridge. Were they all in one tall pile, in several short piles, or lined up across the bridge?
3. What do you think would happen if you spread the pennies out along the beam?
4. What happened when you placed pennies on the arch bridge? [The bridge didn't sag as much as the beam bridge.]
5. What was the difference in the design of the two bridges?
6. Which bridge seemed to be stronger? Why do you think this?
7. Since we used the same materials for each bridge and the same pennies to test both bridges, can we say that it was the difference in the design or shape that made one bridge stronger than another? Explain.

Extensions
1. Share other pictures of bridges with the students and encourage them to build models of these examples.
2. Allow the students to test different types of materials as well as designs for their bridges.

Hold the Load

With 10

Measurement

Starting Measurement

Bridge Builder

With 10

Measurement

Starting Measurement

BUILDING BRIDGES

Topic
Technology: Design

Key Question
How does the design of bridges built from index cards affect the strength of the structure?

Focus
Students will design bridges made of index cards. Through their exploration they should discover that different designs work better than others.

Guiding Documents
Project 2061 Benchmarks
* *People may not be able to actually make or do everything that they can design.*
* *When parts are put together, they can do things that they couldn't do by themselves.*

NRC Standards
* *Scientists use different kinds of investigations depending on the questions they are trying to answer. Types of investigations include describing objects, events, and organisms; classifying them; and doing a fair test (experimenting).*
* *Abilities of Technological Design:*
  * *Identify a simple problem.*
  * *Propose a solution.*
  * *Implement proposed solutions.*
  * *Evaluate a product or design.*
  * *Communicate a problem, design, and solution.*

NCTM Standards
* *Understand the attributes of length, capacity, weight, area, volume, time, temperature, and angle*
* *Develop the process of measuring and concepts related to units of measurement*

Math
Measurement
    length
    mass

Science
Technology
    design

Integrated Processes
Observing
Comparing and contrasting
Communicating
Collecting and recording data
Applying

Materials
For the class:
    the story *The Three Billy Goats Gruff*
    one set of the bridge *Study Prints* (from *Exploring Bridges*)
    reference books (see *Curriculum Correlation*)
    Yes and *No* recording labels (see *Management 3*)
    class chart (see *Management 3*)

For each student:
    three 3″ x 5″ index cards
    clay (see *Management 1*)
    transparent tape *(optional)*
    balance
    one *River Crossing* page (see *Management 2*)
    10 Teddy Bear Counters (see *Management 5*)

Background Information
    The students will build a bridge using index cards to support different masses of clay. If the students have been given prior experiences with this type of building, it is suggested that this activity be used as an assessment in design and problem solving.

Management
1. To give the students experience in using the balance, give each student more clay than is needed to form a ball equal in mass to two Teddy Bear Counters. For *Part 2* each student will need clay equal in mass to 16 Teddy Bear Counters.
2. Duplicate the *River Crossing* page for each student.
3. Enlarge and extend the *Did Your Bridge Support …* class chart and the *Yes* and *No* recording labels to accommodate all the students in the class. The chart should be made large enough to hold the index card models. Place the *Yes* and *No* cards next to the *Did Your Bridge Support …* class chart. The students will choose the appropriate cards to respond to the chart.
4. For *Part One*, students will look at pictures of bridges to become acquainted with various types.

5. The Teddy Bear Counters can be substituted with any non-customary unit such as Unifix cubes, small blocks, etc.

Procedure

Part One
1. Review the bridge *Study Prints* and share several reference books with pictures of bridges (see *Curriculum Correlation*).
2. Compare and contrast these different types of bridges.
3. In small groups, challenge each student to use three 3″ x 5″ cards to build a bridge. Do not give any direction at this point. Allow the students to explore the materials on their own.
4. Give each student some clay. Direct them to form the clay in a ball equal to the mass of two Teddy Bear Counters. Tell them to use the balance to measure the mass of the clay ball.
5. To test their bridge for strength, direct them to place the ball of clay on top of their bridge.
6. Some of the bridges will hold the clay and some will not. Discuss the differences in design between the bridges that held the clay and the ones that did not.
7. Allow the students to try building their bridges again.
8. Be careful to simply offer suggestions and not to show the students a step-by-step procedure to build the bridges. You may want to suggest that the students try folding the cards, but do not show them how. (This will add strength to the card material.) The students will discover many different designs for their bridges.
9. Encourage the students to share with each other by describing the different designs. Tell them to describe how and why they made their bridge the way they did.

Part Two
1. Review the story *The Three Billy Goats Gruff* with the class.
2. Discuss what they have learned about bridge building.
3. Give each student clay, a balance, and Teddy Bear Counters.
4. Direct the students to form three balls of clay. The first ball will need to have a mass of ten Teddy Bear Counters. This ball will represent the big Billy Goat Gruff. Their second ball of clay with a mass of five Teddy Bear Counters will represent the second Billy Goat Gruff and the third ball of clay should have a mass of one Teddy Bear Counter which will represent the smallest Billy Goat Gruff.
5. Challenge the students to design and build a bridge that spans the river on the *River Crossing* page while supporting the big Billy Goat Gruff. You will need to set a specific size that the bridge will need to span. For example, the bridge needs to span the river at a point where it is four Unifix cubes across.

6. Instruct the students to use Unifix cubes to measure several points of the river until they find an appropriate crossing point. (The *River Crossing* page is designed for four different spans of bridges: three, four, five, and six Unifix cubes across.)
7. Allow plenty of building and rebuilding time.
8. If their bridges will not support the big Billy Goat Gruff, suggest they try the smaller goats or a different bridge design.
9. Direct the students to draw a picture or glue their final bridge design to another piece of paper and to mount this illustration or model on the class, *Did Your Bridge Support …* chart. Direct them to glue *Yes* or *No* recording labels in the appropriate spaces on the chart across from their illustrations or models.

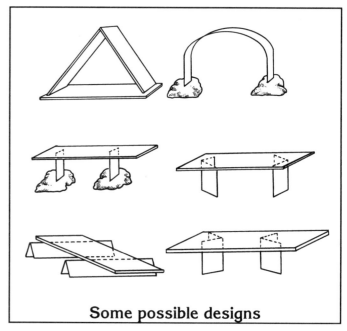

Some possible designs

Discussion
1. Why do we use bridges?
2. What bridge design seemed to work best for the big Billy Goat Gruff?
3. What bridge design could hold all three goats at the same time?
4. What did you do to make your bridge strong?
5. What materials did you use to build your bridge?
6. What tools did you use to build your bridge?
7. Why do you think your bridge was or wasn't strong enough for the clay goats to stand on?
8. How can you make your bridge stronger?
9. What did you do to make your index cards strong?
10. When you had to make your bridge go across a 4 Unifix cubes wide river, what did you have to change on your bridge? … 5 cubes wide? … 6 cubes wide?

Extensions
1. Allow the students to use other types of materials to build their bridges.
2. Leave the materials out for an extended time for the students to further explore bridges.

3. To convey the idea that bridges are used to move lots of different things from one place to another, allow students to try other objects on top of their bridges such as toy cars, trains, action-figure people, etc.

Curriculum Correlation
Literature

Brown, David J. *How Things Were Built*. Random House. NY. 1992.

Brown, Marcia. *The Three Billy Goats Gruff*. Harcourt Brace & Company. NY. 1957.

Edom, Helen. *How Things Are Built*. Usborne Publishing Ltd. London. 1989.

Gaff, Jackie. *Building Bridges and Tunnels*. Scholastic, Inc. NY. 1991.

Galdone, Paul. *The Three Billy Goats Gruff*. Clarion Books. NY. 1973.

Mitgutsch, Ali. *From Cement to Bridge*. Carolrhoda Books. Minneapolis. 1981.

Robbins, Ken. *Bridges*. Dial Books. NY. 1991.

Sheppard, Jeff. *I Know a Bridge*. Macmillan. NY. 1993.

Spier, Peter. *London Bridge is Falling Down!* Doubleday. NY. 1967.

Stevens, Janet. *The Three Billy Goats Gruff*. Harcourt Brace & Company. NY. 1987.

Building Bridges

YES	NO	YES	NO
YES	NO	YES	NO
YES	NO	YES	NO
YES	NO	YES	NO
YES	NO	YES	NO

Recording Labels

Building Bridges

Did Your Bridge Support …

A picture of your bridge	Big Billy Goat	All 3 Billy Goats	Some Billy Goats

BUILDING BRIDGES
River Crossing

HUFF 'n PUFF HOUSES

Topic
Technology: Design/Materials

Key Question
What kind of materials and design do you need to use to construct a building that can withstand the push of the *Big Bad Hair Dryer*?

Focus
Through the design and testing of various model buildings, students will discover that some materials and designs make sturdier structures.

Guiding Documents
Project 2061 Benchmarks
- *Tools are used to do things better or more easily and to do some things that could not otherwise be done at all. In technology, tools are used to observe, measure, and make things.*
- *Tools are used to help make things, and some things cannot be made at all without tools. Each kind of tool has a special purpose.*
- *A model of something is different from the real thing but can be used to learn something about the real thing.*
- *Use hammers, screwdrivers, clamps, rulers, and scissors.*
- *Make something out of paper, cardboard, wood, plastic, metal, or existing objects that can actually be used to perform a task.*

NRC Standard
- *Abilities of Technological Design:*
  - *Identify a simple problem.*
  - *Propose a solution.*
  - *Implement proposed solutions.*
  - *Evaluate the design.*
  - *Communicate the problem, design and solution.*

NCTM Standard
- *Make and use estimates of measurement*

Math
Measurement
Charting

Science
Technology
    materials
    design

Physical science
    forces
    wind energy

Integrated Processes
Observing
Predicting
Comparing and contrasting
Collecting and recording data

Materials
For the class:
    an electric hair dryer (blow dryer)
    a variety of materials to use for building the houses
      (see *Management 3*)
    3" x 5" note cards
    measuring tools, optional (see *Management 4*)
    glue
    tape
    sticky dots or stamps

Background Information
Problem solving is used throughout this activity which finds its origins in the story *The Three Little Pigs*. The students are presented with a challenge as architects and builders to design and construct a house that will withstand the force of a wind from the *Big Bad Hair Dryer*.

Students will need to create and illustrate a design for their houses. They will need to decide upon the construction materials (and quantities) to be used. Once these decisions have been made, they will build and test the houses using the blow dryer. The students will collect and analyze data to try to determine characteristics of houses that are able to withstand the wind test. As always, the students will be encouraged to improve upon their initial design.

In any building activity, the materials used are very important. The physical properties of these materials will determine, to a great extent, the size and strength of the structure. Wood, paper, clay, drinking straws, cardboard, and plastic all present different challenges. When children are encouraged to work freely on a building project, the materials themselves do much of the "teaching."

Management
1. Provide some materials in a building center for free exploration before introducing this activity to your students.

2. In *Part One* of this activity, students are taken through a step-by-step process to build a house out of seven 3″ x 5″ cards. This will help them be more successful when they create a design and choose their own materials in *Part Two*.

3. For *Part Two*, provide a variety of materials from which the students can choose. Some suggestions are: 3″ x 5″ note cards, craft sticks, drinking straws, pipe-cleaners, paper, cardboard, grass and straw, sticks, Styrofoam, blocks of wood, etc.

4. In *Part Two*, the students will design and build their own houses of a determined size. Choose an option according to the ability level of your students: The house should be the "same size" as the house built in *Part 1*; or the house should be 13 cm tall (5″), 8 cm (3″) deep, and 8 cm (3″) wide, have a roof, and be enclosed on all sides. If your students are not using customary units of measurements, you may wish to give them a piece of string for each of the above measurements and tell them their house must be as tall, deep, and wide as the sample strings.

Procedure

Part One

1. Read the story of *The Three Little Pigs* to the class. Encourage the students to use the building center to design their own pig houses.

2. After a period of free exploration in the building center, review the story of *The Three Little Pigs*. Discuss the different designs and materials the students have been using in the building center to build their houses. Relate this to the different materials used by the pigs in the story.

3. Inform the students that they will be making pig houses out of 3″ x 5″ cards. (See *House Construction Instructions*.)

4. After the houses are built, ask the students to describe how they built their houses, what problems they had, etc. Encourage them to think about how they could make their houses better if they were to build them again.

5. Invite the students to predict whether their houses will survive the huffing and puffing of the *Big Bad Wolf*. Tell them that there is not a wolf hanging around, but that you do have a *Big Bad Hair Dryer!*

6. Test each house by setting it on a table, turning the hair dryer on, and pointing it at the house's front door. If the house does not move from its original position, it has survived! If it moves, it has not survived the Big Bad Hair Dryer!

7. If the houses do not survive, ask for suggestions of how they could be changed so they will survive. Suggestions might include putting rocks inside the house, putting craft sticks around the outside to make them sturdier, etc.

Part Two

1. Ask the students if they can think of materials the pigs could have used to build their houses that would survive the huffing and puffing of the Big Bad Hair Dryer.

2. Tell them that they are going to be architects that design and build houses. Advise them that their houses should be able to survive the force from the wind of the Big Bad Hair Dryer.

3. Inform the students about the size of the house.

4. Distribute the *Blueprints* page and inform the students that on the top half of the page, they will need to draw the house they want to build. On the bottom half, they will need to write, illustrate, or glue small bits of the materials and quantities of materials they will need.

5. Show the students the hair dryer to remind them that their houses will need to be designed to survive the huffing and puffing from this Big Bad Hair Dryer!

6. Give the students other items such as string, glue, clay, and tape to help them hold the houses together.

7. Once the students feel their material lists and designs are complete, direct them to gather the materials they will be using and begin building.

8. Allow ample time for the students to work on their houses. Tell them that if they need to make changes in their designs, they need to also make those changes on their drawings. Instruct them to get a new page of *Blueprints* so they can keep a record of their changes.

9. Once the houses are complete, have the students show them, describe the materials used, and explain their construction process.

10. Show the students where to display their houses along with the folded prediction sheet *Will my house survive?* Allow each student to go on a "house tour" and direct them to place a sticky dot or stamp in the appropriate response section (*Yes* or *No*) on the prediction sheet.

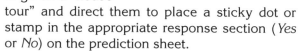

11. Turn on the hair dryer and test the houses. If the house does not move, it has survived.

12. As soon as the tests are complete, direct the students to write their name (*Architect*) and the name of the person holding the hair dryer (*Building Inspector*) in the blanks on the *Did my house survive?* Have them mark their house as either *Yes* or *No*.

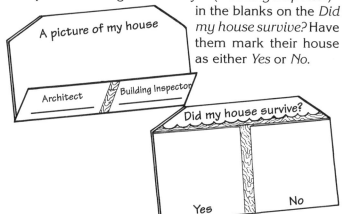

13. Be sure to provide the opportunity to redesign and rebuild their houses, and try the test again.

Discussion

1. What type of material did you choose to use to build your house? Why?
2. Explain how you built your house.
3. Why do you think your house did or did not survive the Big Bad Hair Dryer?
4. When we look at the houses that survived the Big Bad Hair Dryer, which materials were best for this test?
5. For those houses that did not survive, do you think a different shape would have worked better? Explain.
6. What did you learn when you built this house?
7. How are the houses that survived alike? How are they different?
8. How are the houses that did not survive alike? How are they different?
9. What do you think you could do differently next time to make your house better?
10. What tools did you use to build your house?
11. Do you think people build houses the same all over the world? Do you think they use the same materials? Explain.

Curriculum Correlation

Literature:

Gibbons, Gail. *How a House is Built.* Holiday House. NY. 1990.

Hogner, Franz. *From Blueprint to House.* Carolrhoda Books. Minneapolis. 1986.

Klinting, Lars. *Bruno the Carpenter.* Henry Holt and Company. NY. 1995.

Extension

Let students use Lincoln Logs®, Tinker Toys®, or LEGO® elements to create new buildings.

Home Links

1. Ask the students to build other houses at home and to bring them to school to test them against the Big Bad Hair Dryer.
2. Ask students to illustrate their own house and to share it with the class.

House Construction Instructions

Materials for each house:
  seven 3" x 5" note cards
  transparent tape

Procedure:

1. Lay four note cards side by side on the table. Tape their seams.

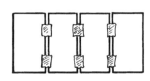

2. Bring the cards together to form a box and tape the open sides together.

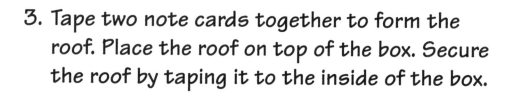

3. Tape two note cards together to form the roof. Place the roof on top of the box. Secure the roof by taping it to the inside of the box.

4. Make one-inch folds of both ends of the last note card. Fit the folded card onto the bottom of the box and tape in place.

5. Draw a front door and some windows on the note card house.

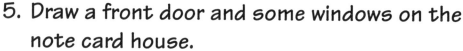

Blueprints

Architect _____

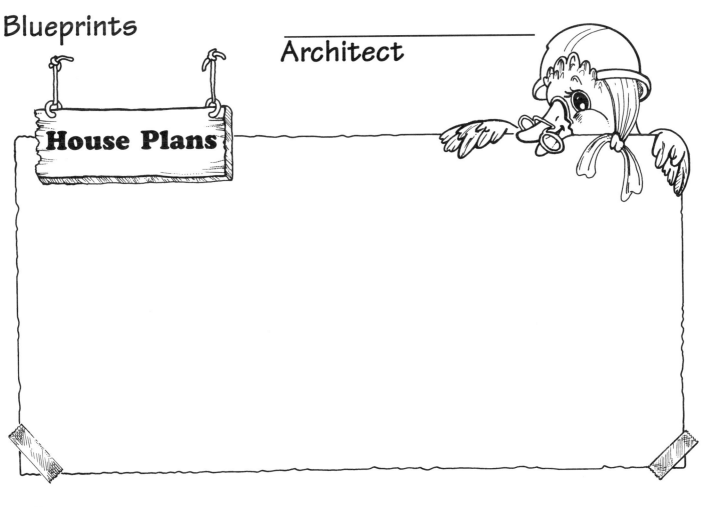

House Plans

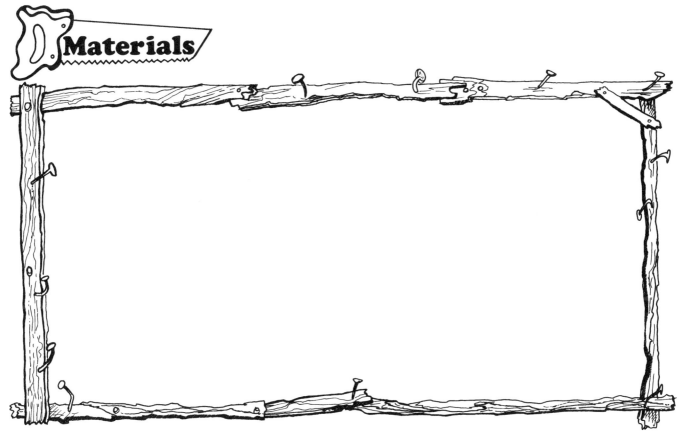

Materials

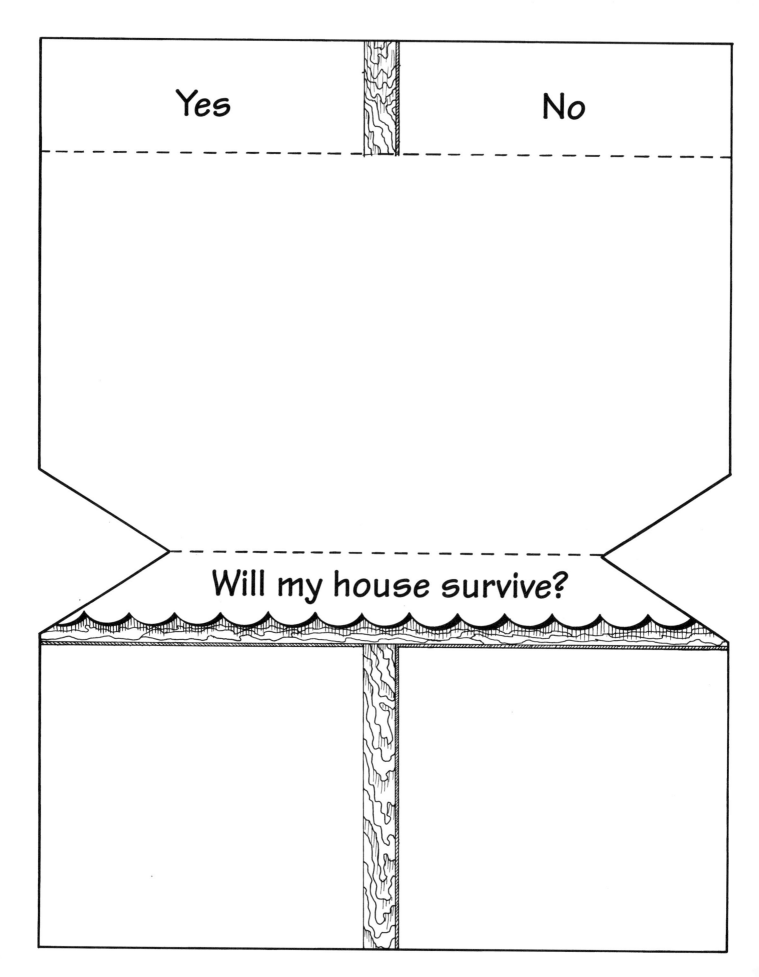

Yes

No

Will my house survive?

98

Architect

Building Inspector

No

Yes

Did my house survive?

A picture of my house

Size and Scale

SIZING UP BEARS

Topic
Technology: Design/Scale

Key Question
When something seems small, is it always small?

Focus
The students will understand that size is a relative concept. The descriptive size of an object depends on the size of the other objects to which it is compared.

Guiding Documents
Project 2061 Benchmark
- *Describing things as accurately as possible is important in science because it enables people to compare their observations with those of others.*

NRC Standard
- *Objects have many observable properties, including size, weight, shape, and color. Those properties can be measured using tools, such as rulers and balances.*

NCTM Standards
- *Develop number sense*
- *Collect, organize, and describe data*
- *Relate physical materials, pictures, and diagrams to mathematical ideas*
- *Make and use estimates of measurement*

Math
Measurement
Ordinal numbers
Counting
Equalities and inequalities
    size comparison

Science
Technology
    scale in design

Integrated Processes
Observing
Comparing and contrasting
Communicating
Applying

Materials
For each student:
    stuffed Teddy Bear
    set of paper Teddy Bears
    one strip of adding machine tape, approximately one
        meter long

Background Information
When in the process of designing, students must consider scale (size comparisons). Young learners are not experienced enough with the magnitude of numbers normally involved with teaching scale, however, they can make simple comparisons between what is smaller or larger.

Measurement is the process of making comparisons between what is being measured and a standard. The standard can be another object or a non-customary or customary unit. A key to success in measurement is multiple experiences with first-hand practice of comparing. For this activity the students will be using qualitative comparisons only. The actual process of measuring will be dealt with in activities to follow.

In this activity the students match and compare objects without the use of formal tools of measurement. They will order these objects using comparison strategies.

Management
1. Ask students to bring in a Teddy Bear. If you are concerned that some students will not have a bear, be sure to have extras on hand to loan to those students.
2. To either introduce or to culminate this activity, you may want to choose one or more of the suggested literature books to read to your class. Each book deals with relative size and will serve as another media to introduce this concept to your class.
3. Duplicate a set of paper Teddy Bears for each student.
4. Cut a meter-length strip of adding machine tape for each student.

Procedure
Part One
1. Read the story of *Goldilocks and the Three Bears* to the class. Lead the students in a discussion about the sizes of the three bears in comparison to each other and to Goldilocks.

102

2. Gather the children with their Teddy Bears in a circle.
3. Allow time for the students to talk about their bears — why the bear is special to them, sharing a description of their bears, etc.
4. Ask that anyone who brought a small bear to stand. Direct those students to gather together in a group. Then group the middle-sized bears in a separate area and the same for the large bears. Be sure to allow the students to make their own judgments as to whether their bear is small, medium, or large.
5. Seat the students in their groups in an arrangement so they can still observe the other groups and their bears.
6. Choose a bear from the small-sized bear group and put it in the middle of the class.
7. Ask someone from the small bear group to place a bear that is larger than the first bear in the circle.
8. Discuss with the students that even though the second bear is coming from the group of students who felt that their bears were small, that it can now be called larger when compared to the smaller bear chosen.
9. Ask someone else from the small bear group to place a bear that is smaller than the first bear in the circle.
10. Ask a student to show the rest of the class the bear in the circle that is large, the bear that is medium, and the bear that is small.
11. Do the same with other bears from each of the middle-sized and large bear groups.
12. Discuss how the descriptive size of an object depends on what they are comparing it to.
13. Once you have demonstrated the comparison procedure with all three groups, allow each group to continue to compare other bears in a similar fashion.

Part Two
1. Gather the students into one large group. Ask them to try to identify the smallest bear in the class. Direct the child holding this bear to place their bear in front of the room.
2. Ask the students to take turns placing their bears next to this bear in order from the smallest to the largest bear in one long line.
3. Instruct one student to choose a bear. Tell them to choose two other bears that would fit into the following sequence:
   This bear is small compared to that bear.
   This bear is large compared to that bear.
4. Choose other students to select a different bear and to continue with the sequence.
5. To practice ordinal counting, ask the students to describe the bear that is second in the line, fourth in line, etc. comparing the size of each one to the sizes of its neighbors. For example, the fourth bear in the line is larger compared to the first bear in line, but smaller compared to the seventh bear in line.

Part Three
1. Distribute a set of paper Teddy Bears to each student.
2. To record their understanding, direct the students to cut out and then glue the paper bears in order from the smallest to the largest onto a strip of adding machine tape.

Discussion
1. How do you decide if something is small, medium, or large? [I compare it to other objects.]
2. Show me something that is small compared to you. Show me something large compared to you. Do the same for your bear.
3. Name something larger than you. Name something smaller than you. Do the same for your bear.
4. Find something that is the same size as you. Do the same for your bear.
5. If we were going to design clothes or furniture for these bears, what do their sizes have to do with our project? Explain.
6. Explain how size affects the design of objects.

Extensions
1. Instead of Teddy Bears, use leaves, shoes, a variety of stuffed animals, or structures they have built in class.
2. Do this activity using non-customary units to quantify the measurement and compare the heights of each bear.
3. Do this activity using customary units of measurement to compare each bear.
4. Do the activity *Fit For A Bear* in which students construct objects which are made to scale for their bears.

Curriculum Correlation
Literature
Alborough, Jez. *Where's My Teddy?* Candlewick Press. Cambridge, MA. 1992. A little boy loses his little Teddy Bear in the forest. A great big bear loses his great big Teddy Bear as well. Each find the other's Teddy Bear and realize that the boy's Teddy Bear is too small for the bear and the bear's Teddy Bear is too large for the boy.

West, Colin. *Hello, Great Big Bullfrog!* Harper & Row, Publishers. NY. 1987. A bullfrog feels great big until he meets a great big rat who meets a great big warthog who meets a great big .... Finally the frog begins to feel quite small in comparison to all the other animals until he meets a great big bumble bee and feels big again.

Ziefert, Harriet. *How Big Is Big?* Puffin Books. NY. 1989. The author compares different objects to one another in relation to size from elephants and whales to insects and planets.

105

MITTS for KITS

Topic
Technology: Design/Materials/Tools/Scale

Key Question
How does the length of a clothesline needed to dry mittens change according to the size of the mittens?

Focus
Students will build a clothesline for the three kittens' mittens. The length of the clothesline will depend on the size of the mittens. The students will decide on the best design, materials, and tools to use for their clothesline.

Guiding Documents
Project 2061 Benchmark
- *In doing science, it is often helpful to work with a team and to share findings with others. All team members should reach their own individual conclusions, however, about what the findings mean.*

NCTM Standards
- *Understand the attributes of length, capacity, weight, area, volume, time, temperature, and angle*
- *Make and use estimates of measurement*
- *Make and use measurements in problem and everyday situations*
- *Develop number sense*

Math
Number sense
Measurement
Equalities and inequalities

Science
Technology
    design
    materials
    tools

Integrated Processes
Observing
Contrasting and comparing

Sorting and classifying
Communicating
Generalizing

Materials
For each student:
    one pair of mittens (see *Management 1*)

For each group of students:
    three sets of paper mittens (see *Management 2*)
    a variety of materials (see *Management 3*)
    Unifix cubes (see *Management 4*)
    scissors
    paper clips, one per mitten

For the class:
    story of *The Three Little Kittens*

Background Information
The activity begins with students comparing objects. They will compare and contrast the lengths and widths of mittens in a classroom collection. Introduce such vocabulary as: longer than, shorter than, as long as, the same as, wider than, as wide as.

This activity combines the use of measurement skills with problem solving. In *Part Two*, each group will be given a set of mittens — all groups will have the same size mittens. They will need to determine the length of clothesline on which to hang these mittens. In *Part Three*, groups will be given sets of mittens that are of a different size than the first set. They can look at the similarities or differences in the length of their clothesline with a group that may or may not have a similar set of mittens. They can then make relative comparisons of as long as, longer than, shorter than, the same as, etc.

Management
1. Prior to teaching this activity, ask the students to bring in a pair of mittens from home.
2. In *Part Two*, all groups of three students will use a set of six paper mittens of the same size. In *Part Three*, groups will be given a set of six paper mittens of a different size than their first set. There are three different sizes of mittens provided for *Part Three*. Each group can compare and contrast their particular sized mittens with the sets of a different size.
3. Supply a variety of materials that can be used to build clotheslines. For example: rulers, dowels, pencils, clay, string, paper towel tubes, paper clips, etc.

4. Each group of students will need a minimum of 25 Unifix cubes.

5. Depending on the developmental level of your students, you may want to precut the mittens for them.

6. Students can use the paper clips to hang their mittens on the clothesline.

Procedure
Part One
1. In groups of three, direct the students to combine the mittens they brought from home. Tell them to sort these mittens according to color, size, shape, etc. Ask them to share their rules for sorting with the rest of the class.

2. Direct them to lay their mittens down in a line. Tell them to measure their line of mittens using Unifix cubes or some other type of non-customary unit.

3. Have students compare their measurements with other groups. Discuss how different-sized mittens take up more or less space.

4. Direct the students to find another pair of mittens that are longer than their pair, shorter than their pair, and the same length as their pair.

5. Discuss the differences in the sizes and shapes of the mittens.

Part Two
1. Read the story *The Three Little Kittens*.

2. Discuss with the children the events in the story bringing the focus to the drying of the mittens. Discuss different ways the kittens could have dried their mittens after washing them. (Many students will suggest a clothes dryer.) If no one responds with using a clothesline, make that suggestion. Then ask the students how long the clothesline needs to be to dry all the kittens' mittens.

3. Ask the students how many mittens there would be if each kitten were to wear mittens on only its front paws. [6 mittens]

4. Once the class has determined the number of mittens needed for three kittens, distribute six paper mittens (all the same size) to each group of three students.

5. Show the students the variety of materials made available to use in building their clotheslines.

6. Challenge them to use these materials to construct a clothesline which will hold all their kittens' mittens. Explain that their clotheslines will need to be strong enough to hold all the mittens.

7. Distribute Unifix cubes to each group. Tell them that these can be used as a measuring tool. Urge the students to determine the needed length of clothesline, how to support the line, etc. This is meant to be a free exploration time for the students to use a trial and error approach to solving these problems. If necessary, encourage students to use a Unifix cube train to help them measure. (They may need to cut the string

a little bit longer than the Unifix cube train to allow some length for attaching the clothesline poles.)

8. Once the clotheslines are complete, ask the students to explain how they determined how long to make their clotheslines. Tell them to explain how they measured the mittens and the string.

9. Compare and contrast the different lengths of clothesline used for the six mittens. Discuss these findings. Ask the students to include the lengths in Unifix cube units.

Part Three
1. Show the students the sets of larger mittens. Have them compare the sizes of these mittens with the mittens already on the line. [They are longer, wider, etc.] Ask what would happen if they wanted to hang these new mittens on the clotheslines they just built. [They wouldn't fit; they're too big.]

2. Give each group a set of six of these new, larger mittens. (All six mittens in each set need to be the same size. Different groups will have different-sized sets.) Challenge each group to build a clothesline for the new set of mittens. Again, encourage the use of Unifix cubes as a measuring tool.

Discussion
1. How did you decide how long to make your clothesline?
2. What did you use to build your clothesline?
3. Do all your mittens fit on your clothesline?
4. Is there room for any more mittens on your clothesline?
5. What was the hardest part to figure out about building your clothesline?
6. If you were going to build another clothesline, what would you do differently?
7. Why is it important to know how to build the right-sized clothesline?
8. If your mittens were the same size as the mittens of another group, would your clothesline work for theirs? Why or why not?
9. Can you find another clothesline in the room that your mittens would not all fit on? … would fit on?
10. What did you learn about building a clothesline?
11. How does the length of your Unifix cube train compare to another group's? What does that mean?

Extensions
1. Direct the students to moisten their second set of mittens using either an eyedropper filled with water or a spray bottle. Ask the students to observe what happens to their clothesline when more weight is added with the water. Direct the students to improve their design by adding supports where needed.

2. Allow students to build another clothesline using a different size set of mittens.

3. Allow students to design a clothesline to dry different articles of clothing.

Part 1—Mitten Sets

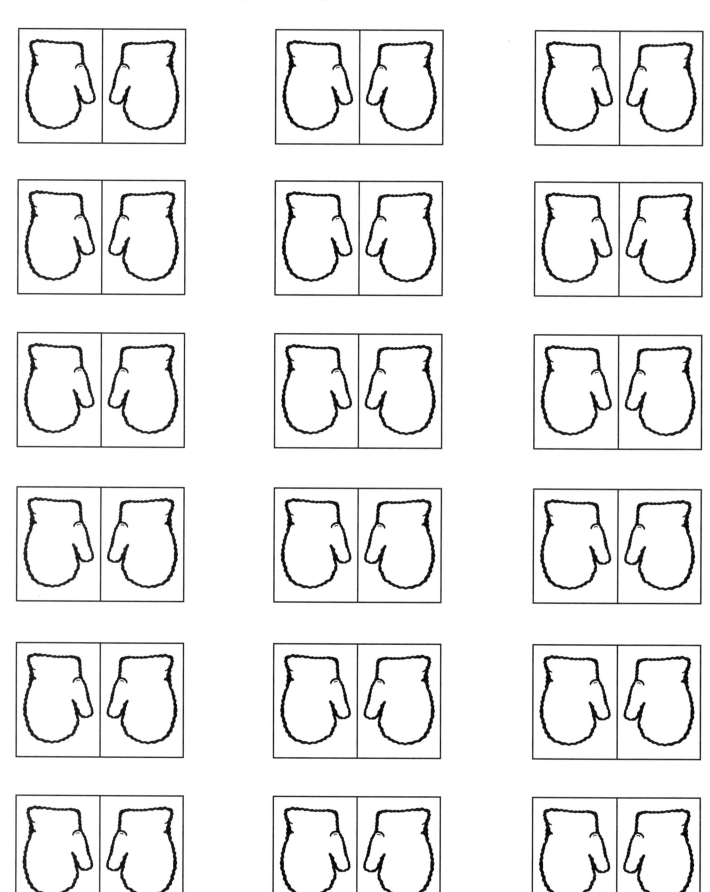

Part 2—Mitten Sets

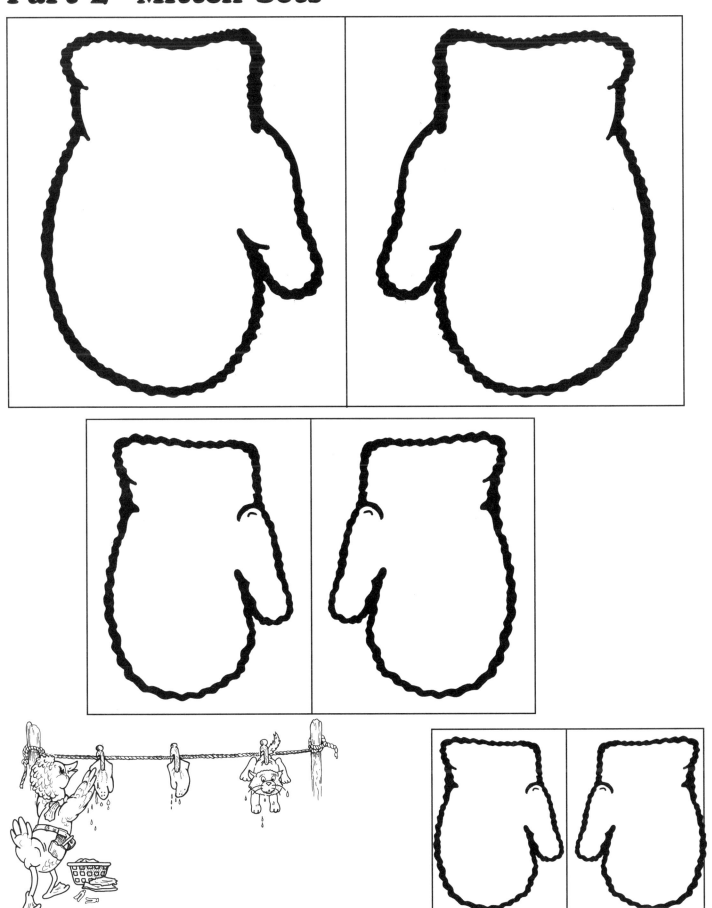

FIT for a BEAR

Topic
Technology: Materials/Tools/Design/Scale

Key Question
How do you know when something is too big, too small, or just right?

Focus
Students will use objects related to themselves and their stuffed animals to construct an understanding of *too big*, *too small*, and *just right*. They will demonstrate their understanding through the design and construction of objects.

Guiding Documents
Project 2061 Benchmarks
• *Tools are used to do things better or more easily and to do some things that could not otherwise be done at all. In technology, tools are used to observe, measure, and make things.*
• *Tools are used to help make things, and some things cannot be made at all without tools. Each kind of tool has a special purpose.*
• *Use hammers, screwdrivers, clamps, rulers, scissors, and hand lenses, and operate ordinary audio equipment.*
• *Make something out of paper, cardboard, wood, plastic, metal, or existing objects that can actually be used to perform a task.*

NCTM Standards
• *Use mathematics in other curriculum areas*
• *Relate physical materials, pictures, and diagrams to mathematical ideas*
• *Make and use measurements in problem and everyday situations*

Math
Number sense
Measurement
Ordering
Equalities and inequalities

Science
Technology
    materials
    tools
    design

Integrated Processes
Observing
Comparing and contrasting
Recording

Materials
For the class:
    a collection of adult-, child-, and infant-sized clothing
    a large chair, a small chair, and one that would be considered just right for most students in your class
    various-sized shoe boxes
    construction paper
    scraps of fabric
    safety pins
    Velcro® (optional)
    glue
    scraps of wood
    hammers
    nails
    transparent tape
    scissors
    Measuring Tape (see *Management, Part 2, No. 3*)

For each group of four students:
    a Teddy Bear or stuffed animal from home

Background Information
Students will experience the feel of trying on clothes that are too big, too small, and just right. After several concrete experiences, they will use their own stuffed animals to demonstrate their understanding of relative size by building objects which are too big, too small, and just right as related to the story *Goldilocks and the Three Bears*. Through the process of building, the students will begin to apply their knowledge of materials, tools and design to their projects.

In early years, students develop skills such as cutting, connecting, switching on/off, pouring, holding, tying, and hooking. They can use non-customary units to make linear measurements of objects and materials. They can begin to understand the mechanics of construction through the use of blocks of wood, hammers, nails, cardboard boxes and glue. Through the actual construction of objects they have personally designed, students can begin to develop an understanding of strength of materials, needed support for structures, and size. Problem-solving techniques are developed as they face obstacles along the way.

It is important that students understand that measurement is never exact, that even the most careful measurements are approximations. Children need to learn to evaluate when their measurements are "close enough." Children also need practice in making estimates in measurement.

Management

In *Part One,* students will find things that are too big, too small, and just right for them. In *Part Two,* they will construct some clothes, a chair, and a bed that are just right for their stuffed animals. They will compare the sizes of these items with other groups to determine too big, too small, and just right. *Part Three* can be used to record their understanding of relative size.

Part One
1. Bring in adult-, child-, and infant-sized clothes, shoes, etc. You might want to place these in a home or drama center for a couple of weeks prior to introducing this activity.
2. Demonstrate to the students the proper fit of a chair to a person. The person should be able to sit all the way back in the chair with his/her back resting against the back of the chair. The feet should rest flat on the floor and the knees should be level with the chair seat.

Part Two
1. Ask the students to bring in their favorite Teddy Bear or stuffed animal. One stuffed animal per group is needed.
2. Collect several different-sized shoe boxes for building beds; construction paper or scraps of cloth for clothing; safety pins, glue or Velcro® to serve as fasteners and to make seams; pieces of wood to construct a chair; hammer and nails; tape; thread; etc. (You may want to send a letter home requesting many of the supplies.)
3. Have several different types of measuring tools available to the students, customary and non-customary. Duplicate and assemble the *Measuring Tape* for each group of students.

Part Three
1. Duplicate one set of the paper Teddy Bears, clothing, and chairs for each student.
2. Duplicate a journal for each student.

Procedure
Part One
1. Read *Goldilocks and the Three Bears* and/or other suggested literature books to the students (*see Curriculum Correlation*). Discuss what it means for something to be too big, too small, and just right.

2. Place a collection of clothes in front of the students. Ask a student to find something that is too big and to put it on over his/her school clothes.
3. Have students explain how they know that this is too big.
4. Assign the same task to another student.
5. Follow *Procedures 2-4* with students finding clothes that are too small and those that fit just right. Be sure to have students communicate how they know the clothing fits these categories.
6. Ask a student to find a chair in the classroom that is too big and explain why that chair was chosen.
7. Follow the same procedure as above with a chair that is too small and one that is just right.

Part Two
1. Bring all the Teddy Bears/stuffed animals to the center of the group. Ask the students to arrange them in order from the smallest to the largest.
2. Guide a discussion as to whether these Teddy Bears would wear the same size clothes or use the same size chair.
3. Tell the students that each group is to make three things for a Teddy Bear: some clothes, a chair, and a bed that is just right for their bear.
4. Distribute one Teddy Bear for each group of four students. Tell them to gather the materials they will be needing from the construction station and to begin working.
5. Upon completion, ask each group to share their bear and constructed pieces and to discuss why they feel their objects are just right for their bear. Ask them to look at the other groups' items and to find some clothes that would be too big and some that would be too small for their bear. Continue with the other objects in the same manner.
6. Invite them to tell the rest of the class how they built each piece, what they used, and how they put it together.
7. Direct the students to line the chairs up in order from the largest to the smallest, the beds in the same type of line, and then the bears in the same type of line. Direct the students to compare the sizes of the bears with the sizes of the constructed pieces. Have them determine if the bears match up with the chairs and beds that were constructed for them. Help the students make the connection that the size of the bear determines the chair or bed or clothes that are just right for that bear.

Part Three
1. Distribute the paper Teddy Bears, clothing, and chair pages. Direct the students to cut out the items. Have them dress one of their bears with the clothing which will fit just right and to place the bear in the chair that is just right.
2. Give each student a journal. Once the students have shown you their *just right* example, have them glue it to the *Just Right* journal page.
3. Ask students to dress one of the remaining bears in clothing that is too small and place that bear in

a chair that is too small. This will be glued on the *Too Small* journal page.

4. The remaining bear should be dressed in clothing which is too big and placed in a chair that is too big. This should be glued on the *Too Big* journal page.

5. Urge the students to illustrate the front cover of their journals.

Discussion

1. How do you know when something is too big, too little, or just right?

2. How did you decide how big or how little to make the things for your bear? Did you measure anything? Explain.

3. Tell how you made your objects. What materials did you use to build each one? What did you use to keep it together? What tools did you use?

4. What did you consider (think about) before deciding on the design of the bed? ... the chair? ... the clothes?

5. What did you learn about building things in this activity?

6. What did you like to build the most? Why?

7. If you were going to build another set of objects for your bear, what would you do differently? ... the same?

8. Were you able to build everything you wanted to? Did you think of some designs that you were not able to build? Explain.

Extension

Have a *Too Big Day* and ask all the students to wear one of their parent's shirts to school.

Curriculum Correlation

Literature

Myller, Rolf. *How Big Is A Foot?* Dell Publishing. NY. 1990. Measurement is a focus in this story. The king asks to have a bed designed and made for the queen. The need for measurement is demonstrated after a failure of communication results in a bed that does not fit the queen.

Roberts, Tom. *Goldilocks.* Rabbit Ears Books. Westport, Connecticut. 1990. This famous folk tale refers to chairs and beds that are too big, too small, and just right for Goldilocks.

Turkle, Brinton. *Deep In The Forest.* E.P. Dutton. NY. 1976. This story is another version of *Goldilocks and the Three Bears*, but turned around. This time a bear visits the cabin of a human family.

Tolhurst, Marilyn. *Somebody and the Three Blairs.* Orchard Books. NY. 1990. A spin-off book of *Goldilocks and the Three Bears*.

Home Link

Ask the students to bring in a piece of clothing that is too big for them and one that is too small for them.

FIT for a BEAR

FIT for a BEAR

114

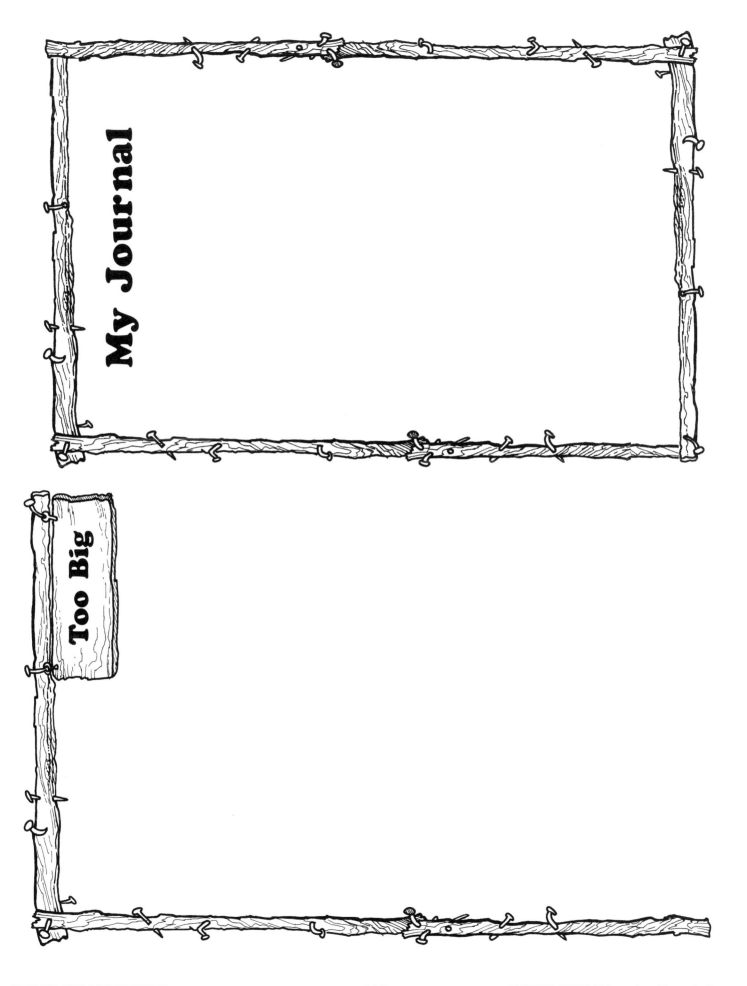

My Journal

Too Big

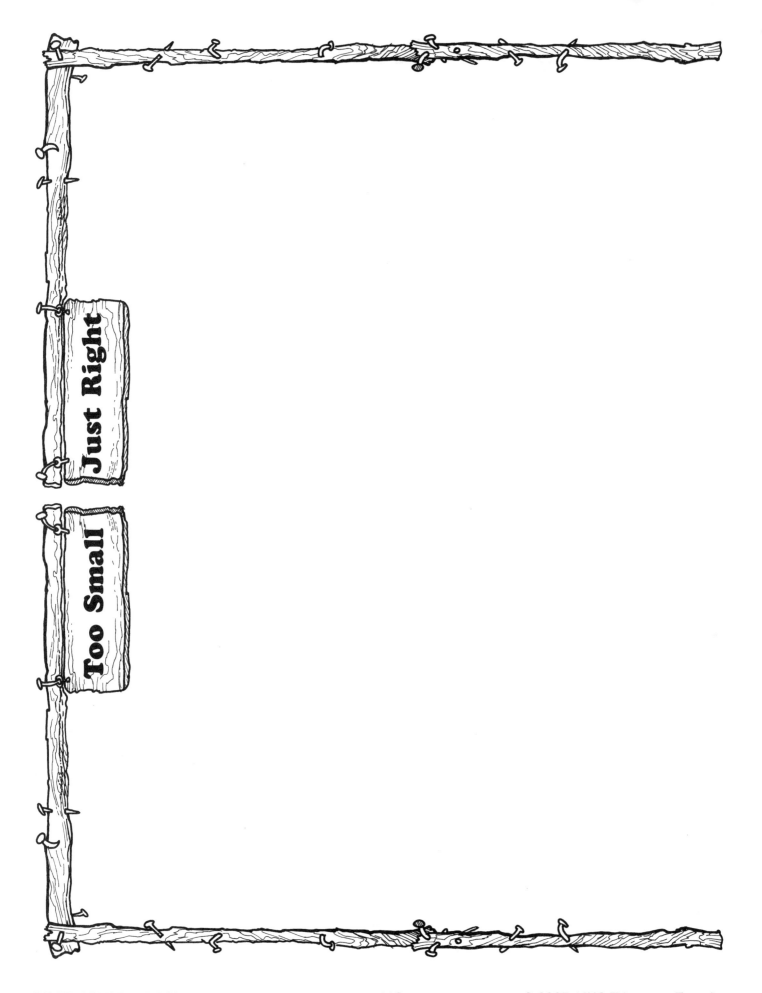

Just Right

Too Small

FIT for a BEAR
Measuring Tape

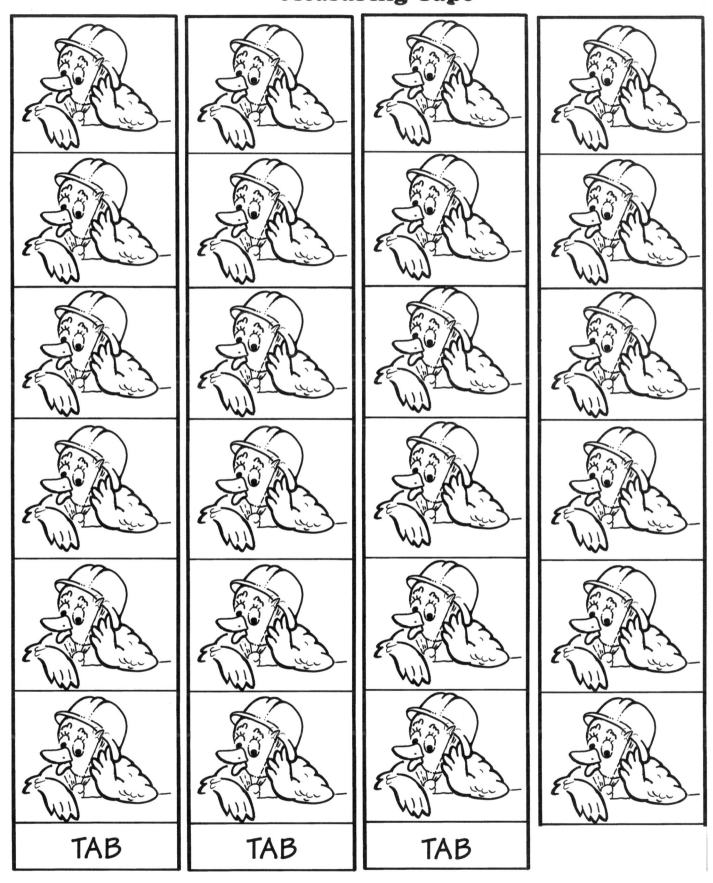

TAB   TAB   TAB

Design on the Mind

An inventor I am,
'Cause I love to design,
So I'm thinking about
This new project of mine.

I need to make plans—
It just has to be good.
Should I make it of metal,
Or paper or wood?

What material is best?
What will help it be strong?
Just how wide should it be,
And how tall and how long?

Let me think, will it work?
Can it move and bend too?
Can it do all the things
That I want it to do?

Is it just the right size?
Does it need nails or glue?
Will it do the job well?
Will it last till it's through?

What support does it need,
And what shape should it be?
What will make it look great
For my buddies to see?

Yes, I love to design—
It's a whole lot of fun.
And I can't wait to see
What I have when I'm done!

—Suzy Gazlay

Technology Projects

Pull-It Eggs

Topic
Technology: Design

Key Questions
How does the string make the puppet's arms and legs move at the same time?
What happens when you change the design of the puppet?

Focus
Students will observe the reaction of a pull-string puppet when its string is pulled. They will also discover that by changing the design of the puppet, the movement of the puppet changes.

Guiding Documents
Project 2061 Benchmarks
- *When trying to build something or get something to work better, it usually helps to follow directions if there are any or to ask someone who has done it before for suggestions.*
- *The way to change how something is moving is to give it a push or a pull.*

NRC Standards
- *Scientists use different kinds of investigations depending on the questions they are trying to answer. Types of investigations include describing objects, events, and organisms; classifying them; and doing a fair test (experimenting).*
- *The position and motion of objects can be changed by pushing or pulling. The size of the change is related to the strength of the push or pull.*

Science
Technology
   design
Physical science
   force and motion

Integrated Processes
Observing
Comparing and contrasting
Applying
Recording

Materials
For the class:
   several pull-string puppets (see *Management 1*)

For each puppet:
   egg puppet copied onto cardstock
   2 pieces of string, 50 cm (20″) and 25 cm (10″)
   4 paper fasteners
   scissors
   crayons
   hole punch

Background Information
This activity serves as an introduction into the integration of technology projects into the language arts. In this directed activity, students will be exposed to a simple pull-string puppet. They will explore its design and then change it in order to observe the effects of this change.

When the students pull on the string of the first design, the arms and legs of the toy move upward — a study in action/reaction and cause/effect relationships. When they release the string, the arms and legs fall downward. After the students have explored how this design works, they will be encouraged to alter the design and observe what effects the changes have on the puppet.

NOTE: This activity requires fine motor skills which some young learners may not have developed. It can be taught using teacher-made puppets.

Management
1. You can find commercial puppets like the one in this activity. They are often sold as Christmas ornaments.
2. If you are using this lesson with very young learners, prepare at least one puppet per group of students. If you are working with older students, allow the students to make their own puppets. Follow the pictured directions on the student page for construction of the puppet.
3. Copy the *Pull-It Egg* onto cardstock.
4. Do not use a coarse quality string because it will cause so much friction going through the holes that the puppet's arms and legs will not move freely.

Procedure

1. Show the students a commercial pull-string puppet or the egg puppet they will be using. Demonstrate its action/reaction by pulling on the string. Ask them to describe how they think this pull-string puppet might be working.

2. Show the puppet with its arms and legs at rest.

3. Using different objects around the room, show that when you pull on something, that object usually moves or changes in some way. Tell students that pushes and pulls are called *forces*. Ask students to apply some push or pull forces to objects in the room. With each student demonstration, discuss whether it is a push or a pull.

4. Bring the students' attention back to the puppet and ask what they observe happening when the string is pulled then released. Make sure that in the discussion it is brought out that when a *force* is applied to an object, that object may move or change in some way.

5. Inform the students that they are going to use their own puppet to observe the results of pull forces.

6. Distribute the pre-made samples of the pull-string puppet. Tell the students that they are to work as scientists to observe and test the different things their puppet can do. Direct them to carefully pull on the strings to make arms and legs move.

7. Ask them to retell the *Humpty Dumpty* nursery rhyme and when they get to "Humpty Dumpty had a great fall," have them pull the string to raise the puppet's arms and legs.

8. Point out to the students that this puppet was designed by humans and not by nature. Explain that they can also design and build objects that can be used to entertain us.

9. Tell the students that a scientist sometimes tests the object they are exploring by making a change in its design. Ask the students to try changing the placement of the paper fasteners. For example: move them to a different hole on the arms or legs. Then tell them to try pulling on the strings again and to observe the puppet's movement, comparing and contrasting it to the original movement.

10. Allow the students to make different changes to the puppets and to observe and discuss the results.

Discussion

1. What makes the puppet's arms and legs move? [pulling the string]

2. Why do the arms and legs move up when you pull down on the string?

3. What else is moving besides the arms and legs of your toy? [my hand is moving, the string is moving]

4. What happened when you attached the string in a different position?

5. Describe what happened when you changed the position of the paper fasteners. Did this change the materials, the tools, or the design?

6. Describe what you did as a scientist today.

Extensions

1. Change the egg to a spring rabbit, the Gingerbread Man, Santa Claus, an animal, a child, etc. Adapt to other characters in the literature studies of the class.

2. Challenge the students to design their own puppets.

Home Link

Ask the students to bring in other pull-string puppets they may have at home to share with the rest of the class.

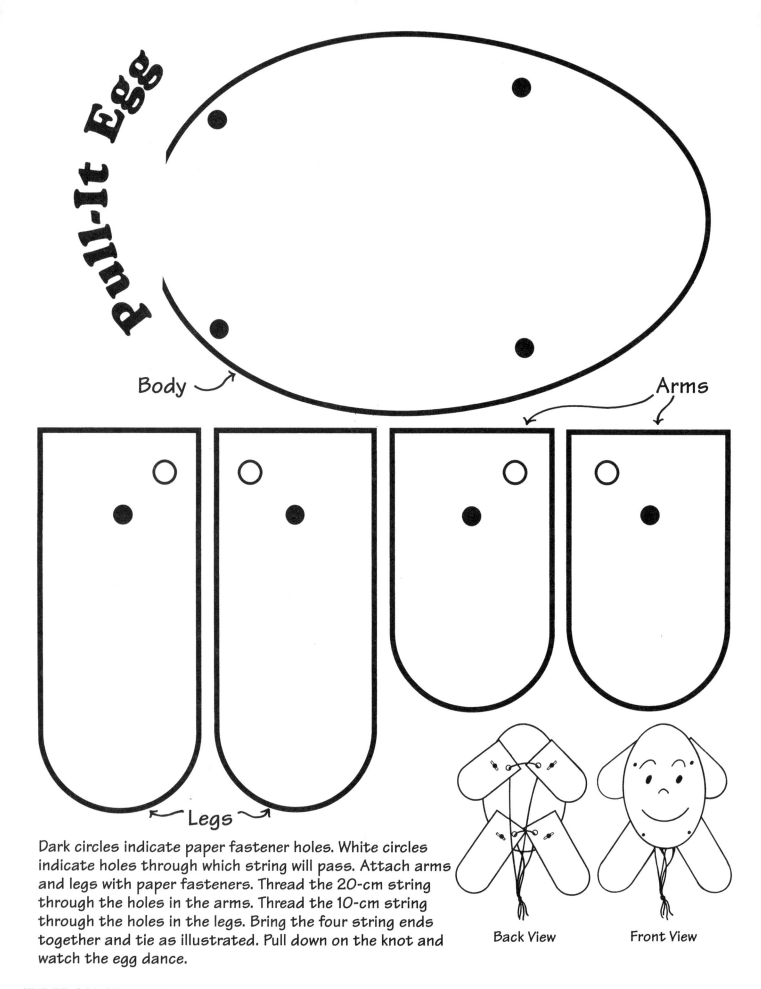

Pull-It Egg

Body

Arms

Legs

Dark circles indicate paper fastener holes. White circles indicate holes through which string will pass. Attach arms and legs with paper fasteners. Thread the 20-cm string through the holes in the arms. Thread the 10-cm string through the holes in the legs. Bring the four string ends together and tie as illustrated. Pull down on the knot and watch the egg dance.

Back View

Front View

PUSH 'n PULL Puppets

Topic
Technology: Design/Materials/Tools

Key Question
How can we design our puppets so they can be made to move in different ways?

Focus
Students will apply their understanding of design, materials, and tools to building puppets.

Guiding Documents
Project 2061 Benchmarks
- *When trying to build something or get something to work better, it usually helps to follow directions if there are any or to ask someone who has done it before for suggestions.*
- *Things move in many different ways, such as straight, zigzag, round and round, back and forth, and fast and slow.*
- *Several steps are usually involved in making things.*
- *When parts are put together, they can do things that they couldn't do by themselves.*
- *Make something out of paper, cardboard, wood, plastic, metal, or existing objects that can actually be used to perform a task.*

NRC Standard
- *Communicate investigations and explanations.*

NCTM Standards
- *Formulate problems from everyday and mathematical situations*
- *Make and use measurements in problem and everyday situations*

Science
Technology
   design, tools, materials
Physical science
   force and motion

Integrated Processes
Observing
Recording
Comparing and contrasting
Communicating
Applying

Materials
For each student:
   scissors
   tape
   glue
   variety of building materials (see *Management 2*)
      paper fasteners, 1″ and 1/2″
      coffee stirrers or straws
      7-oz waxed paper cups
      6″ paper plates
      1″ - 2″ diameter balls (see *Management 3*)
      spools
      craft sticks
      fabric scraps (see *Management 3*)
      wood scraps
      cardboard
      cylinders (paper tubes)
      tube socks
      bobby pins (see *Management 6*)
      rubber bands
      yarn
      buttons
      felt
      string
   variety of art materials for decorating puppets

For the class:
   examples of a variety of puppets

Background Information
By designing and building puppets, young learners will apply their understanding of materials, tools, and design to situations tied to literature, nursery rhymes and fairy tales. This child-centered activity allows students to explore the joining of materials in specific ways in order to make their puppets move. After they have been given exploration time, they are then asked to apply what they have learned to their own creations.

Management
1. Gather examples of several different types of puppets. Try to include jointed puppets, string puppets, sock puppets, etc.

2. To help save on the expenses, send home the *Family Letter* prior to teaching this lesson in order to gather many of the materials.

3. Collect 1"-2" diameter balls made from Styrofoam, rubber, or plastic. Fabric pieces for the cup puppets should be approximately 24 cm x 24 cm (9" x 9").

4. Drill holes in the craft sticks large enough to insert paper fasteners. These connections will serve as joints for the puppets.

5. Cut the felt tongue and yarn hair for the sock puppet. (See *Sock Puppet Instructions*.)

6. Open a bobby pin and then fold back one end to form a needle.

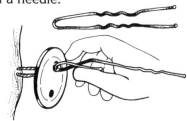

7. Prepare a set of the puppets presented in this activity to use as samples for the students.

8. Plan ample time for exploration. You may want to leave the supplies and puppets at a free-choice center for several weeks.

Procedure
Part One
1. Demonstrate several examples of puppets which move. Use the puppet from *Pull-It Eggs* as an example. Discuss the mechanics of the movement. Give the students the opportunity to work the puppets.

2. Ask the students for suggestions of ways to make puppets.

3. Show them the variety of materials the class has gathered and ask them to offer suggestions on how to use these materials to build a puppet.

4. Using their suggestions, work with the whole class to design and build a puppet. Allow them to implement their ideas and suggestions. This will help give those who have no experience in this type of construction information for their future projects.

5. Once the class puppet is completed, tell the students that they are going to build a puppet of their own. Direct them to design the puppet so that the arms and legs will be able to move in some way.

6. In a small group, allow the students to begin exploring with the materials.

7. Once all students have completed their puppets, allow them to share with the class how they made their puppets and how they move. Encourage the students to discuss the design of the puppets as well as the materials and tools they used.

Part Two
1. Show the students how to make joints using the drilled craft sticks and paper fasteners. Show them how joints allow the arms and legs of the puppet

to bend and change position. Relate these joints to the student's joints in their bodies. Ask them how the puppet's joints are like their joints and how they are different.

2. This puppet is the most difficult design to make. If necessary, build this as a demonstration puppet and allow the students time to explore the movement of the joints. (See *Jointed Puppet Instructions*).

3. Discuss how the design of this puppet allows the puppet to move.

Part Three
1. On another day, show the students puppets made from socks. Point out the parts of the puppets that have been sewn on.

2. Tell them that they will be using materials, tools, and a new design to sew a puppet.

3. In a small group, hand out the tube socks, rubber bands, and felt tongues. Following the directions, guide the students through the process of putting their sock puppet together. (See *Sock Puppet Instructions*.)

4. Once the students have sewn on the hair, button eyes, etc., allow them to again share their puppets, explaining the design, materials, and tools they used.

Part Four
1. The next puppet for the students to build is one using a paper cup, some fabric, a sphere for the head, and craft sticks.

2. Show the students your sample puppet and ask them to describe to you how this puppet is moving. Ask them to describe how they think it was made.

3. In a small group and with adult assistance, guide the students through the process of building the puppet. (See *Cup Puppet Instructions*.)

4. Use the sharing time for the students to explain the design, materials, and tools used.

Discussion
1. Describe how you made one of your puppets.
2. How did it help you to have a design in your mind before you began building?
3. Now that you have built several puppets, what ideas do you have about how you would like to design another puppet? What materials and tools would you need?
4. What did you find out about the materials and tools you used to make your puppets move?
5. How could we use the puppets we built?
6. How could we make our puppets work better?
7. Did you have to measure or count anything when you made your puppet? Explain.
8. What did you learn when you made your puppets?

Extensions
1. Use the puppets in the puppet stage activity at the end of this book.
2. For language arts, drama, etc., use the puppets to retell familiar stories.

124

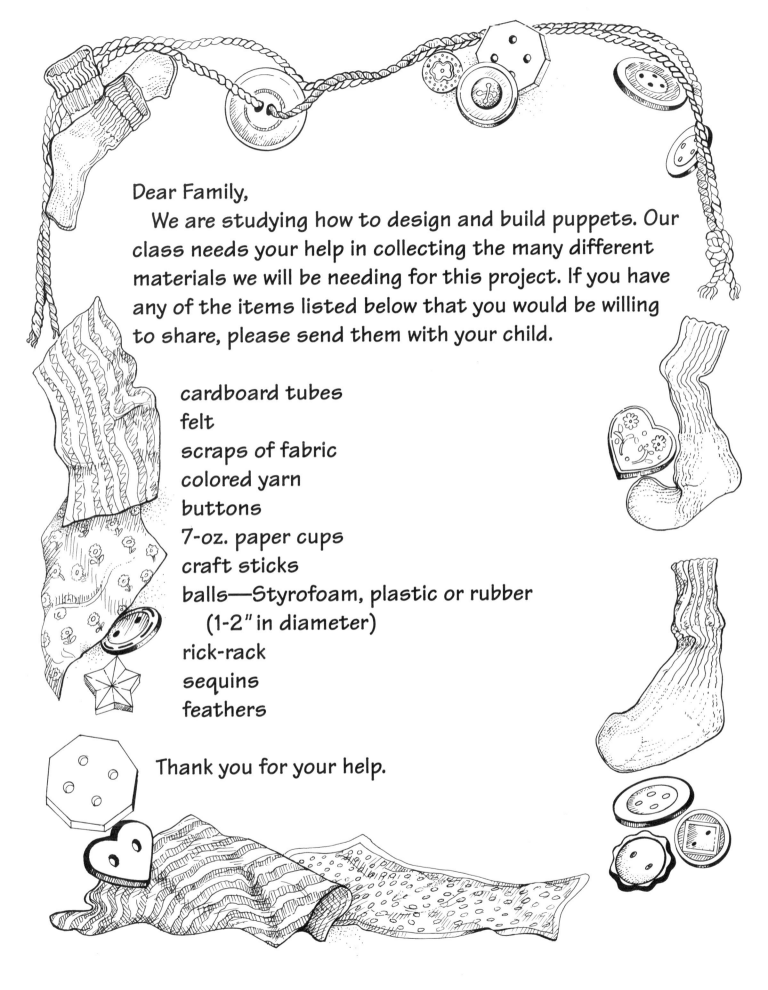

Dear Family,

We are studying how to design and build puppets. Our class needs your help in collecting the many different materials we will be needing for this project. If you have any of the items listed below that you would be willing to share, please send them with your child.

cardboard tubes
felt
scraps of fabric
colored yarn
buttons
7-oz. paper cups
craft sticks
balls—Styrofoam, plastic or rubber
    (1-2" in diameter)
rick-rack
sequins
feathers

Thank you for your help.

Jointed Puppet Instructions

Drill holes into the ends of ten craft sticks.

To make a joint, insert a paper fastener through two holes.

Attach the shoulders, the head, and the legs to the paper cup as illustrated.

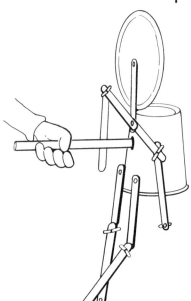

To hold the puppet, push a drinking straw through a hole which has been made in the back of the paper cup.

To animate the puppet, hold onto the straw and lift the puppet up and down.

A more controlled way to animate the puppet is to attach strings to the ends of the two arms and use a cross bar.

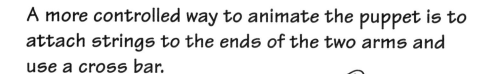
Glue

Sock Puppet Instructions

Use a tube sock for the body of the puppet.

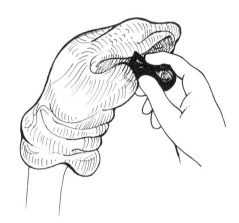

Reach inside the sock and grasp the tongue, pulling it into the inside of the sock. Pull until the sock is inside out.

Use a rubber band to wrap the end of the sock which has the tongue inside.

Turn the sock right-side out.

Using a bobby pin needle and yarn, attach buttons for eyes and a nose.

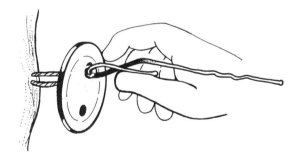

Use the needle to attach the ears and yarn hair.

Add other embellishments as desired.

Sock Puppet Pattern

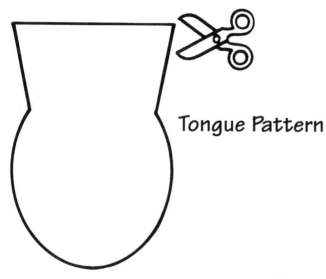

Tongue Pattern

Ear Pattern

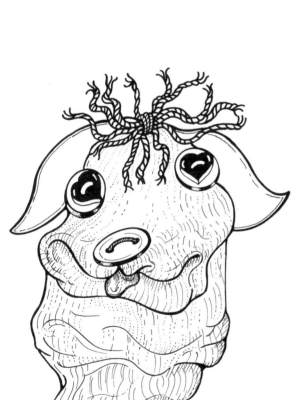

Place broken line on fold and cut on solid line.

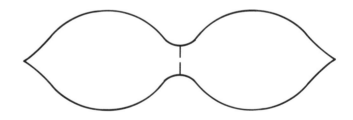

The ears should look like this after they are cut.

Cup Puppet Instructions

Fold a piece of cloth (24 cm x 24 cm or 9" x 9") in half and glue the two side edges together.

Glue the bottom edge inside the rim of a 7-oz paper cup.

To make a handle, use masking tape to join three craft sticks together.

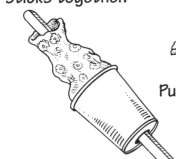

Push the handle up through the bottom of the cup.

Gather the top edges around the tip of the handle and secure with a rubber band.

Poke a hole in the ball that is large enough for the craft stick handle to fit securely inside.

To attach the head, push the ball onto the end of the handle.

Attach flexible arms (designed by the student and cut from paper or felt) to the back of the cloth puppet.

Embellish the face using glue and various craft items or markers.

After the glue has dried, gently pull the handle down to collapse the puppet in the cup.

Carefully pushing the handle up will show the puppet. The puppet can be animated by turning the handle back and forth and all around.

Puppet Stage

Throughout this study of technology, the students have made many models and props which can be incorporated into a theater-arts experience. Direct the students to examine their constructions and then design a plan for a story which they can present to their classmates, parents, etc.

Challenge the students to construct a puppet stage which would be appropriate to use with their puppets, props, and models. Leave this challenge very open-ended, making it the students' responsibility to gather the materials and tools needed for their designs.

Encourage the students to use their constructions throughout the year, refining their techniques, creating new props, writing new stories, etc.

Designer Books

The following designer books integrate the study of technology to the language arts. Technology is addressed through the design and construction of the books. Literacy is accessed through the communication of the stories involved.

Step-by-step instructions are included to help with the initial construction of the books. However, they are intended to be used as a launching pad for the students to design and construct their own projects.

An additional suggested resource for book and display designs is the publication *Big Book of Books and Activities* by Dinah Zike (Dinah-Might Activities, Inc. San Antonio, TX. 1993).

Designer Book 1

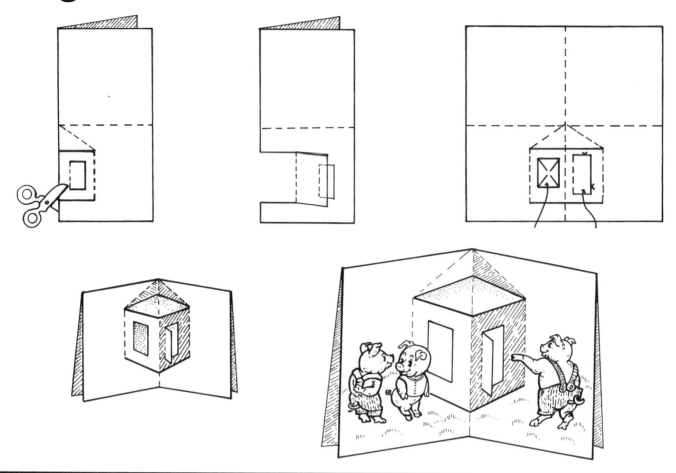

Designer Book 2

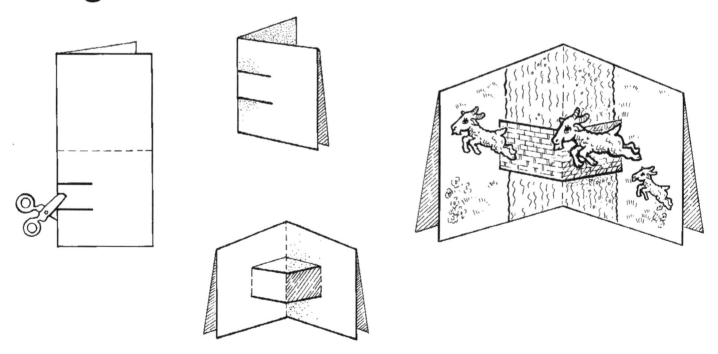

132

Designer Book 3

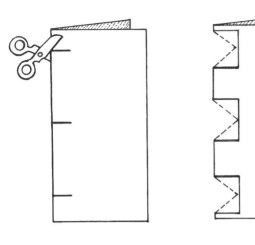

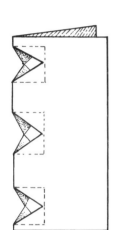
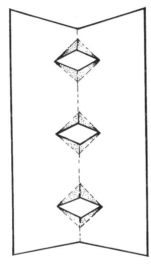

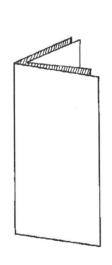

Designer Book 4

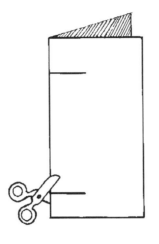
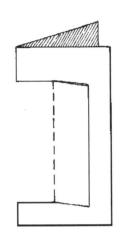

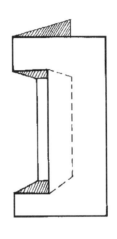
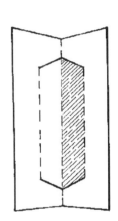

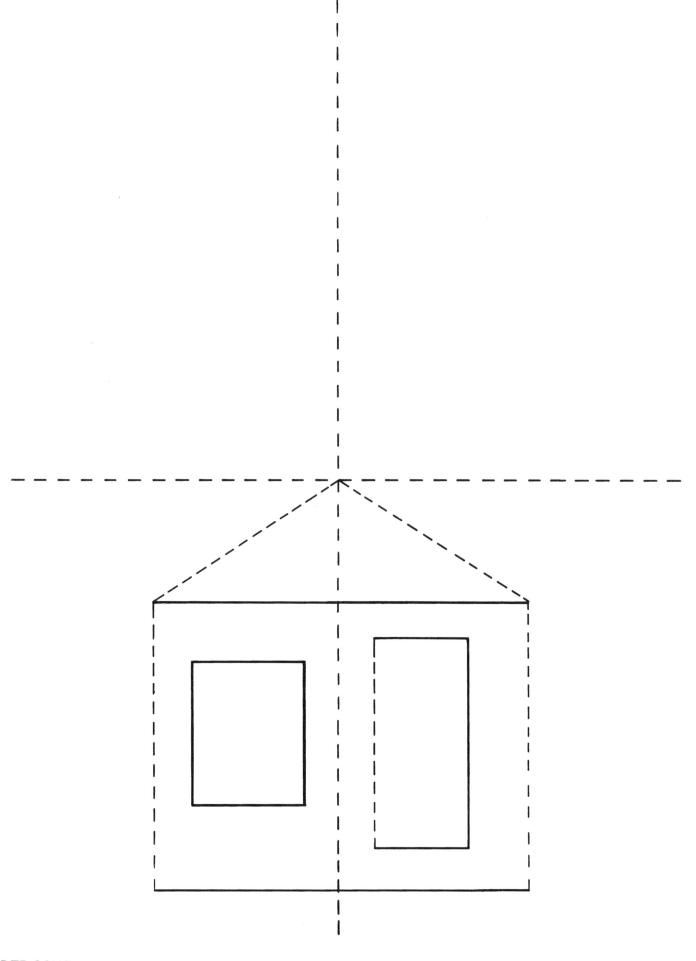

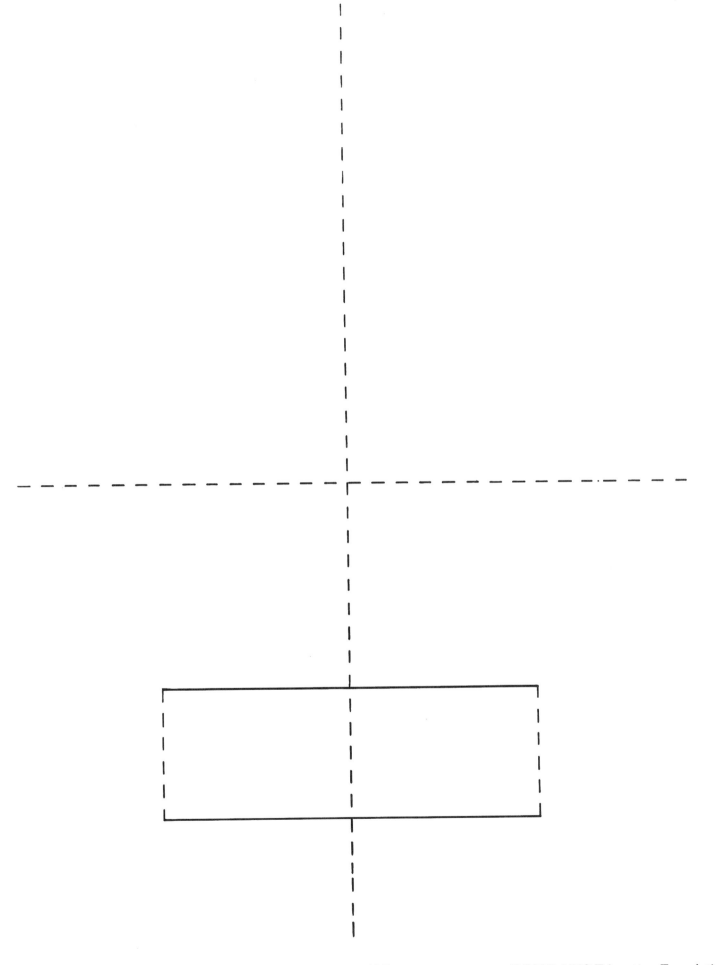

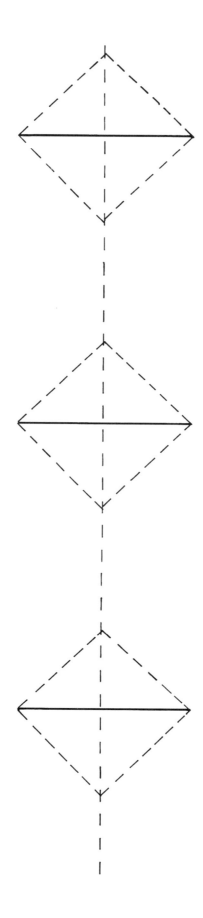

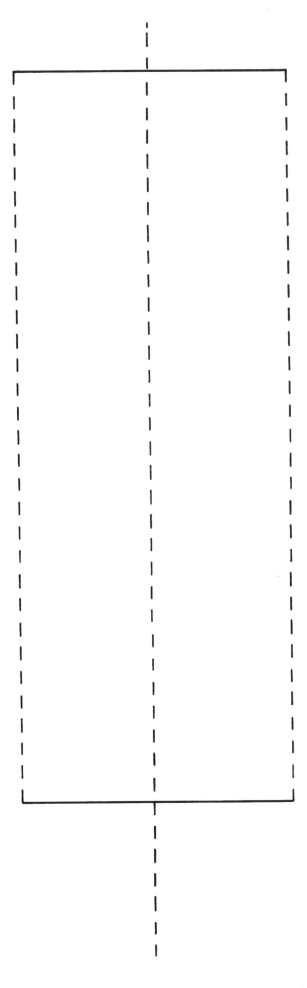

137

Bibliography

Construction

Allen, Judy. *What is a Wall, After All?* Candlewick Press. Cambridge, MA. 1993.

Brown, David J. *How Things Were Built.* Random House. NY. 1992.

Edom, Helen. *How Things Are Built.* Usborne Publishing Ltd. London. 1989.

Gaff, Jackie. *Building Bridges and Tunnels.* Scholastic, Inc. NY. 1991.

Gibbons, Gail. *How A House Is Built.* Holiday House. NY. 1990.

Hoberman, Mary Ann. *A House Is A House For Me.* Penguin Books. NY. 1978.

Hogner, Franz. *From Blueprint to House.* Carolrhoda Books. Minneapolis. 1986.

Kitchen, Bert. *And So They Build.* Candlewick Press. Cambridge, MA. 1993.

Klinging, Lars. *Bruno the Carpenter.* Henry Holt and Company. NY.1995.

Lent, Blair. *Bayberry Bluff.* Houghton Mifflin. Boston. 1987.

Mitgutsch, Ali. *From Cement to Bridge.* Carolrhoda Books. Minneapolis. 1981.

Robbins, Ken. *Bridges.* Dial Books. NY. 1991.

Sheppard, Jeff. *I Know a Bridge.* Macmillan. NY. 1993.

Spier, Peter. *London Bridge is Falling Down!* Doubleday. NY. 1967.

Stevenson, Robert Louis. *Block City.* Puffin Books. NY. 1988.

Size and Scale

Alborough, Jez. *Clothesline.* Candlewick Press. Cambridge, MA. 1993.

Alborough, Jez. *Where's My Teddy?* Candlewick Press. Cambridge, MA. 1992.

Myller, Rolf. *How Big Is A Foot?* Dell Publishing. NY. 1990.

West, Colin. *Hello, Great Big Bullfrog!* Harper & Row, Publishers. NY. 1987.

Ziefert, Harriet. *How Big Is Big?* Puffin Books. NY. 1989.

Folk Tales, Nursery Rhymes, and Other Stories

Brown, Marcia. *The Three Billy Goats Gruff.* Hartcourt Brace & Company. NY. 1957.

Celsi, Teresa. *The Fourth Little Pig.* Steck-Vaughn. Austin, TX. 1992.

Galdone, Paul. *The Three Billy Goats Gruff.* Clarion Books. NY. 1973.

Gander, Father. *Father Gander Nursery Rhymes: The Equal Rhymes Amendment.* Advocacy Press. Santa Barbara, CA. 1985

Gill, Shelley. *Alaska's Three Bears.* Paws IV Publishing Company. Homer, AK. 1990.

Hickey. *Mother Goose & More.* Additions Press. Oakland, CA 1990.

Imai, Miko. *Little Lumpty.* Candlewick Press. Cambridge, MA. 1994.

Lowell, Susan. *The Three Little Javelinas.* Northland Publishing. Flagstaff, AZ. 1992.

Roberts, Tom. *Goldilocks.* Rabbit Ears Books. Westport, CT. 1990.

Scieszka, Jon. *The True Story of The Three Little Pigs.* Penguin Books. NY. 1989.

Smith, Jessie Willcox. *The Jessie Willcox Smith Mother Goose.* Derrydale Books. NY. 1986.

Stevens, Janet. *The Three Billy Goats Gruff.* Harcourt Brace & Company. NY. 1987.

Trivizas, Eugene. *The Three Little Wolves and the Big Bad Pig.* Maxwell Macmillan International. NY. 1993.

Turkle, Brinton. *Deep In The Forest.* E.P. Dutton. NY. 1976.

Tolhurst, Marilyn. *Somebody and the Three Blairs.* Orchard Books. NY. 1990.

Vozar, David. *Yo, Hungry Wolf!* Delacorte Press. NY. 1993.

Multi-Media Resources

Jenkins, Ella. *Nursery Rhymes: Rhyming & Remembering for Young Children & for Older Girls and Boys with Special Language Needs.* (Songs) Smithsonian/Folkways.

From Fiber to Fabric. (Video) The National Cotton Council. 901-274-9030.

Wool Challenge. (Video) Victorian Video Productions. 800-848-0284.

The AIMS Program

AIMS is the acronym for "**A**ctivities **I**ntegrating **M**athematics and **S**cience." Such integration enriches learning and makes it meaningful and holistic. AIMS began as a project of Fresno Pacific University to integrate the study of mathematics and science in grades K-9, but has since expanded to include language arts, social studies, and other disciplines.

AIMS is a continuing program of the non-profit AIMS Education Foundation. It had its inception in a National Science Foundation funded program whose purpose was to explore the effectiveness of integrating mathematics and science. The project directors in cooperation with 80 elementary classroom teachers devoted two years to a thorough field-testing of the results and implications of integration.

The approach met with such positive results that the decision was made to launch a program to create instructional materials incorporating this concept. Despite the fact that thoughtful educators have long recommended an integrative approach, very little appropriate material was available in 1981 when the project began. A series of writing projects have ensued, and today the AIMS Education Foundation is committed to continue the creation of new integrated activities on a permanent basis.

The AIMS program is funded through the sale of books, products, and staff development workshops and through proceeds from the Foundation's endowment. All net income from program and products flows into a trust fund administered by the AIMS Education Foundation. Use of these funds is restricted to support of research, development, and publication of new materials. Writers donate all their rights to the Foundation to support its on-going program. No royalties are paid to the writers.

The rationale for integration lies in the fact that science, mathematics, language arts, social studies, etc., are integrally interwoven in the real world from which it follows that they should be similarly treated in the classroom where we are preparing students to live in that world. Teachers who use the AIMS program give enthusiastic endorsement to the effectiveness of this approach.

Science encompasses the art of questioning, investigating, hypothesizing, discovering, and communicating. Mathematics is the language that provides clarity, objectivity, and understanding. The language arts provide us powerful tools of communication. Many of the major contemporary societal issues stem from advancements in science and must be studied in the context of the social sciences. Therefore, it is timely that all of us take seriously a more holistic mode of educating our students. This goal motivates all who are associated with the AIMS Program. We invite you to join us in this effort.

Meaningful integration of knowledge is a major recommendation coming from the nation's professional science and mathematics associations. The American Association for the Advancement of Science in *Science for All Americans* strongly recommends the integration of mathematics, science, and technology. The National Council of Teachers of Mathematics places strong emphasis on applications of mathematics such as are found in science investigations. AIMS is fully aligned with these recommendations.

Extensive field testing of AIMS investigations confirms these beneficial results:

1. Mathematics becomes more meaningful, hence more useful, when it is applied to situations that interest students.
2. The extent to which science is studied and understood is increased, with a significant economy of time, when mathematics and science are integrated.
3. There is improved quality of learning and retention, supporting the thesis that learning which is meaningful and relevant is more effective.
4. Motivation and involvement are increased dramatically as students investigate real-world situations and participate actively in the process.

We invite you to become part of this classroom teacher movement by using an integrated approach to learning and sharing any suggestions you may have. The AIMS Program welcomes you!

AIMS Education Foundation Programs

Practical proven strategies to improve student achievement

When you host an AIMS workshop for elementary and middle school educators, you will know your teachers are receiving effective usable training they can apply in their classrooms immediately.

Designed for teachers—AIMS Workshops:
- Correlate to your state standards;
- Address key topic areas, including math content, science content, problem solving, and process skills;
- Teach you how to use AIMS' effective hands-on approach;
- Provide practice of activity-based teaching;
- Address classroom management issues, higher-order thinking skills, and materials;
- Give you AIMS resources; and
- Offer college (graduate-level) credits for many courses.

Aligned to district and administrator needs—AIMS workshops offer:
- Flexible scheduling and grade span options;
- Custom (one-, two-, or three-day) workshops to meet specific schedule, topic and grade-span needs;
- Pre-packaged one-day workshops on most major topics—only $3,900 for up to 30 participants (includes all materials and expenses);
- Prepackaged *week-long* workshops (four- or five-day formats) for in-depth math and science training—only $12,300 for up to 30 participants (includes all materials and expenses);
- Sustained staff development, by scheduling workshops throughout the school year and including follow-up and assessment;
- Eligibility for funding under the Eisenhower Act and No Child Left Behind; and
- Affordable professional development—save when you schedule consecutive-day workshops.

University Credit—Correspondence Courses

AIMS offers correspondence courses through a partnership with Fresno Pacific University.
- Convenient distance-learning courses—you study at your own pace and schedule. No computer or Internet access required!

The tuition for each three-semester unit graduate-level course is $264 plus a materials fee.

The AIMS Instructional Leadership Program

This is an AIMS staff-development program seeking to prepare facilitators for leadership roles in science/math education in their home districts or regions. Upon successful completion of the program, trained facilitators become members of the AIMS Instructional Leadership Network, qualified to conduct AIMS workshops, teach AIMS in-service courses for college credit, and serve as AIMS consultants. Intensive training is provided in mathematics, science, process and thinking skills, workshop management, and other relevant topics.

Introducing AIMS Science Core Curriculum

Developed in alignment with your state standards, AIMS' Science Core Curriculum gives students the opportunity to build content knowledge, thinking skills, and fundamental science processes.
- *Each* grade specific module has been developed to extend the AIMS approach to full-year science programs.
- *Each* standards-based module includes math, reading, hands-on investigations, and assessments.

Like all AIMS resources these core modules are able to serve students at all stages of readiness, making these a great value across the grades served in your school.

For current information regarding the programs described above, please complete the following:

Information Request

Please send current information on the items checked:

____ *Basic Information Packet* on AIMS materials ____ Hosting information for AIMS workshops
____ *AIMS Instructional Leadership Program* ____ AIMS Science Core Curriculum

Name _____ Phone _____

Address_____
    Street   City   State   Zip

Magazine

YOUR K-9 MATH AND SCIENCE
CLASSROOM ACTIVITIES RESOURCE

The AIMS Magazine is your source for
standards-based, hands-on math and
science investigations. Each issue is
filled with teacher-friendly, ready-to-use
activities that engage students in
meaningful learning.

- *Four issues each year (fall, winter, spring, and summer).*

Current issue is shipped with all past issues within that volume.

| 1822 | Volume XXII | 2007-2008 | $19.95 |
| 1823 | Volume XXIII | 2008-2009 | $19.95 |

Two-Volume Combination
| M20507 Volumes XXI & XXII | 2006-2008 | $34.95 |
| M20608 Volumes XXII & XXIII | 2007-2009 | $34.95 |

Back Volumes Available
Complete volumes available for purchase:

1802	Volume II	1987-1988	$19.95
1804	Volume IV	1989-1990	$19.95
1805	Volume V	1990-1991	$19.95
1807	Volume VII	1992-1993	$19.95
1808	Volume VIII	1993-1994	$19.95
1809	Volume IX	1994-1995	$19.95
1810	Volume X	1995-1996	$19.95
1811	Volume XI	1996-1997	$19.95
1812	Volume XII	1997-1998	$19.95
1813	Volume XIII	1998-1999	$19.95
1814	Volume XIV	1999-2000	$19.95
1815	Volume XV	2000-2001	$19.95
1816	Volume XVI	2001-2002	$19.95
1817	Volume XVII	2002-2003	$19.95
1818	Volume XVIII	2003-2004	$19.95
1819	Volume XIX	2004-2005	$19.95
1820	Volume XX	2005-2006	$19.95
1821	Volume XXI	2006-2007	$19.95
1822	Volume XXII	2007-2008	$19.95

Volumes II to XIX include 10 issues.

**Call 1.888.733.2467 or
go to www.aimsedu.org**

Subscribe
to the
AIMS Magazine

**$19.95
a year!**

AIMS Magazine
is published four
times a year.

Subscriptions ordered
at any time will receive
all the issues for that
year.

AIMS Online—www.aimsedu.org

To see all that AIMS has to offer, check us out
on the Internet at www.aimsedu.org. At our
website you can search our activities database;
preview and purchase individual AIMS activities;
learn about core curriculum, college courses,
and workshops; buy manipulatives and other
classroom resources; and download free resources
including articles, puzzles, and sample AIMS
activities.

AIMS News
While visiting the AIMS website, sign up for AIMS
News, our FREE e-mail newsletter. You'll get the
latest information on what's new at AIMS including:

- New publications;
- New core curriculum modules; and
- New materials.

Sign up today!

AIMS Program Publications

Actions with Fractions, 4-9
Awesome Addition and Super Subtraction, 2-3
Bats Incredible! 2-4
Brick Layers II, 4-9
Chemistry Matters, 4-7
Counting on Coins, K-2
Cycles of Knowing and Growing, 1-3
Crazy about Cotton, 3-7
Critters, 2-5
Electrical Connections, 4-9
Exploring Environments, K-6
Fabulous Fractions, 3-6
Fall into Math and Science, K-1
Field Detectives, 3-6
Finding Your Bearings, 4-9
Floaters and Sinkers, 5-9
From Head to Toe, 5-9
Fun with Foods, 5-9
Glide into Winter with Math and Science, K-1
Gravity Rules! 5-12
Hardhatting in a Geo-World, 3-5
It's About Time, K-2
It Must Be A Bird, Pre-K-2
Jaw Breakers and Heart Thumpers, 3-5
Looking at Geometry, 6-9
Looking at Lines, 6-9
Machine Shop, 5-9
Magnificent Microworld Adventures, 5-9
Marvelous Multiplication and Dazzling Division, 4-5
Math + Science, A Solution, 5-9
Mostly Magnets, 2-8
Movie Math Mania, 6-9
Multiplication the Algebra Way, 6-8
Off the Wall Science, 3-9
Out of This World, 4-8
Paper Square Geometry:
    The Mathematics of Origami, 5-12
Puzzle Play, 4-8
Pieces and Patterns, 5-9
Popping With Power, 3-5
Positive vs. Negative, 6-9
Primarily Bears, K-6
Primarily Earth, K-3
Primarily Physics, K-3
Primarily Plants, K-3

Problem Solving: Just for the Fun of It! 4-9
Problem Solving: Just for the Fun of It! Book Two, 4-9
Proportional Reasoning, 6-9
Ray's Reflections, 4-8
Sensational Springtime, K-2
Sense-Able Science, K-1
Soap Films and Bubbles, 4-9
Solve It! K-1: Problem-Solving Strategies, K-1
Solve It! 2nd: Problem-Solving Strategies, 2
Solve It! 3rd: Problem-Solving Strategies, 3
Solve It! 4th: Problem-Solving Strategies, 4
Solve It! 5th: Problem-Solving Strategies, 5
Spatial Visualization, 4-9
Spills and Ripples, 5-12
Spring into Math and Science, K-1
The Amazing Circle, 4-9
The Budding Botanist, 3-6
The Sky's the Limit, 5-9
Through the Eyes of the Explorers, 5-9
Under Construction, K-2
Water Precious Water, 2-6
Weather Sense: Temperature, Air Pressure, and Wind, 4-5
Weather Sense: Moisture, 4-5
Winter Wonders, K-2

Spanish Supplements*
Fall Into Math and Science, K-1
Glide Into Winter with Math and Science, K-1
Mostly Magnets, 2-8
Pieces and Patterns, 5-9
Primarily Bears, K-6
Primarily Physics, K-3
Sense-Able Science, K-1
Spring Into Math and Science, K-1

* Spanish supplements are only available as downloads from the
  AIMS website. The supplements contain only the student pages
  in Spanish; you will need the English version of the book for the
  teacher's text.

Spanish Edition
Constructores II: Ingeniería Creativa Con Construcciones
    LEGO® 4-9
    The entire book is written in Spanish. English pages not included.

Other Publications
Historical Connections in Mathematics, Vol. I, 5-9
Historical Connections in Mathematics, Vol. II, 5-9
Historical Connections in Mathematics, Vol. III, 5-9
Mathematicians are People, Too
Mathematicians are People, Too, Vol. II
What's Next, Volume 1, 4-12
What's Next, Volume 2, 4-12
What's Next, Volume 3, 4-12

For further information write to:
AIMS Education Foundation • P.O. Box 8120 • Fresno, California 93747-8120
www.aimsedu.org • 559.255.6396 (fax) • 888.733.2467 (toll free)

Duplication Rights

Standard Duplication Rights

Purchasers of AIMS activities (individually or in books and magazines) may make up to 200 copies of any portion of the purchased activities, provided these copies will be used for educational purposes and only at one school site.

Workshop or conference presenters may make one copy of a purchased activity for each participant, with a limit of five activities per workshop or conference session.

Standard duplication rights apply to activities received at workshops, free sample activities provided by AIMS, and activities received by conference participants.

All copies must bear the AIMS Education Foundation copyright information.

Unlimited Duplication Rights

To ensure compliance with copyright regulations, AIMS users may upgrade from standard to unlimited duplication rights. Such rights permit unlimited duplication of purchased activities (including revisions) for use at a given school site.

Activities received at workshops are eligible for upgrade from standard to unlimited duplication rights.

Free sample activities and activities received as a conference participant are not eligible for upgrade from standard to unlimited duplication rights.

Upgrade Fees

The fees for upgrading from standard to unlimited duplication rights are:
- $5 per activity per site,
- $25 per book per site, and
- $10 per magazine issue per site.

The cost of upgrading is shown in the following examples:
- activity: 5 activities x 5 sites x $5 = $125
- book: 10 books x 5 sites x $25 = $1250
- magazine issue: 1 issue x 5 sites x $10 = $50

Purchasing Unlimited Duplication Rights

To purchase unlimited duplication rights, please provide us the following:
1. The name of the individual responsible for coordinating the purchase of duplication rights.
2. The title of each book, activity, and magazine issue to be covered.
3. The number of school sites and name of each site for which rights are being purchased.
4. Payment (check, purchase order, credit card).

Requested duplication rights are automatically authorized with payment. The individual responsible for coordinating the purchase of duplication rights will be sent a certificate verifying the purchase.

Internet Use

Permission to make AIMS activities available on the Internet is determined on a case-by-case basis.

- P. O. Box 8120, Fresno, CA 93747-8120 -
- aimsed@aimsedu.org - www.aimsedu.org -
- 559.255.6396 (fax) - 888.733.2467 (toll free) -